HOW THE EARTH'S PLATE TECTONIC CYCLE WORKS

CMB Reaction

BY

John H. Carman

authorHOUSE®

AuthorHouse™
1663 Liberty Drive, Suite 200
Bloomington, IN 47403
www.authorhouse.com
Phone: 1-800-839-8640

First published by AuthorHouse 6/29/2010

ISBN: 978-1-4389-7351-7 (e)
ISBN: 978-1-4389-8946-4 (sc)

Library of Congress Control Number: 2009905072

Printed in the United States of America
Bloomington, Indiana

This book is printed on acid-free paper.

ABSTRACT

Existing data and use of a hypothetical ternary model post Stishovite (pSt) -Magnesiowustite (Mw)-Iron(Ir), indicate that the Earth's core could be the convertor end member of the Earth's Plate Tectonic Cycle (EPTC), a new theory. In this new theory the third pivotal end member, the core, is the place where the cycle begins and ends, to begin again. The first pivotal end member of the EPTC, for a three pivotal end member system, is the global MORB end member where new oceanic crust and lithosphere are created. Sea-floor spreading connects it to the second pivotal end member, the subduction end member, where crust and lithosphere disappear to become cold lithospheric-crust complexes descending through the dense mantle toward the Earth's core. When 'complexes' break into the Earth's outer core they are slowed, turned and endothermally ingested by it. Partial melting by it frees lower mantle phases and a core phase while forming metallic liquid plus densified immiscible silicate liquid, of which 17 vol.% reduces the convecting bulk density of the Earth's outer core by 10%. Freed crystalline micro-phases, micrometers to millimeters in size, more or less fill mega-bodies of<83 vol.% metallic liquid and <17 vol.% immiscible silicate liquid, both of centimeters to kilometers in size. Excess core energy starting each cycle comes mainly from irreversible exothermal reactions of micro-phases at numerous unstable phase contacts toward stable equilibrium within and between convecting mega-bodies, only to make contact and react again and . .again. Other sources of exothermal energy come from radioactive silicate liquid and from friction at stable phase contacts during convection. Accumulation of heat from these energy sources tends to expand the outer core as univariant boundary reactions reverse, with +5.0 $cm^3 g^{-1}$ coming from the inner core boundary alone. The outer core's pervasive expansion against the passively resisting strength of the mantle results in explosive ejection of silicate liquid along lines of weakness or at points of weakness of the CMB when it fails, as it must. Superheated actions of this silicate liquid with lower mantle phases result in hybrid, hot and solid domains that ascent as hot basic plumes through a cooler and denser ultrabasic mantle. Decompression melting of plumes, above 290 km, contributes new basaltic crust at mid-oceanic ridges, the MORB end member of the EPTC, from curtain like-plumes from line sources of the CMB whereas, new crust for oceanic island basalts comes from point source plumes of the CMB by similar processes through the lower and upper mantle. Both of these basalt types are thought to be linked with silicate liquid ejected from the Earth's outer core as expressed by the ubiquitous yet unexplained C-component of Pb isotope ratios for Atlantic-, Indian- and Pacific mid-oceanic basalts, mayor segments of the MORB end member of the EPTC, and by its corresponding FOZO isotope component of oceanic island basalts. Disappearance of oceanic crust and lithosphere at the subduction end member of the EPTC before it is a thirtieth of the age of the Earth leads to its ingestion and eventual energization in the Earth's outer core, the convertor- and third pivotal end member of the EPTC, to drive ensuing cycles. It seems unavoidable that egress features for silicate liquid on the CMB could serve as a template for the EPTC, as it is for global 'hot-spots', in a bottom up-dynamic, except for the location and actions of the subdcuction end member. There is no shortage of silicate liquid in the outer core, 17 vol.% is almost twice the volume of the entire crust of the Earth. Finally, it seems possible that some of the unexplained heat flux of the Earth, over and above that explained by mantle radioactivity may be found here in this new paradigm that drives the Earth's plate tectonic cycle.

I also conclude that the use of the ternary model and existing data indicates that the current core-mantle boundary is most likely represented by the quaternary univariant reaction,

Mg Perovskite + Ca Perovskite + Magnesiowustite = Iron Alloy Liquid + Silicate Liquid at about 4500°K and 130 to 135 GPa, where Silicate Liquid is peritectic containing ferric iron. And, similar quaternary reactions will be found for the Earth's inner-core boundary and the lower-mantle boundary at the top of the D" layer. This quaternary system would have more than six crucial micro-phases and a hundred and thirty-five binary contacts in each of three or more parts of the outer core, of which one third are stable for friction, and the rest unstable for phase contact reactions. Exothermal energy comes from irreversible liquid -buffered binary, -ternary and -quaternary phase contact reactions of micro-phases toward stable equilibrium with six hundred or so outcomes inside and between convecting immiscible liquid mega-bodies, plus radioactivity of silicate liquid, and friction at stable phase contacts of micro- phases inside and between convecting mega-bodies. These actions and reactions are believed to supply adequate internal energy for a creditably dynamic and variable magnetic field in strength, focus and polarity plus adequate energy to 'drive' the Earth's plate tectonic cycle.

INTRODUCTION

Harry H. Hess (1) gave us the first fundamental change to our early view of 'continental drift' toward what is now known as the Earth's plate tectonics, when he suggested that continents diverge passively away from the mid-oceanic ridges 'riding on mantle material in divergent convection cells' (the 'Sea-floor Spreading Stage' of crust and lithospheric plates moving away from the MORB End Member of Fig. 1and 10) as ocean basins are created and enlarged, rather then by their plowing through a strong mantle due to the influence of gravitational and rotational forces of Alfred Wegener (2). As a balance for a constant volume of the Earth, 'ocean basins are, elsewhere at the same time destroyed', he proposed, 'where oceanic crust and mantle converge in the downward limbs of whole mantle convection cells' (The Subduction End Member where plates of lithosphere and 'Crust Disappears' Fig. 1 and 10). And by way of completion, I indicate that somewhere in the Earth's mantle at one of its density discontinuities, at the D" layer, or in the Earth's core itself, descending lithospheric mantle and oceanic crust is converted (The Convertor End Member where ' Descending Cold Feeds are Converted into Ascending Hot Feeds Fig. 1 and 10) by heat and caused to ascend to create 'new oceanic basalt crust' and lithosphere at the mid-oceanic ridge and rise (MORB End Member Fig. 1 where 'oceanic crust is created'). Plates then diverge to be subducted before they are l/30th of the age of the Earth. The place and processes of this conversion are the foci of this paper, in order to take the plate tectonics to a more complete theory of the Earth's plate tectonic cycle.

The Gibbs Phase Rule (3) is used here as an application of the tenents of thermodynamics to explore the hypothesis that the Earth's core is the location of the convertor end member of the EPTC. This is accomplished by a review of phase relations bearing on the lower mantle and core; secondly, by forming an internally consistent set chemical components to represent the phase relations for a simple hypothetical model ternary system under hypersolidus to subsolidus conditions for the Earth's lower mantle and core; thirdly, by using the hypothetical model to suggest univariant reactions for the lower-mantle (LMB) (for the top of the D" layer), for the core-mantle boundary (CMB), for the bottom of the D" Layer, and for the inner-core boundary (ICB); and fourthly, we see this model's actions work now, and toward the ultimate complete crystallization of the Earth's core.

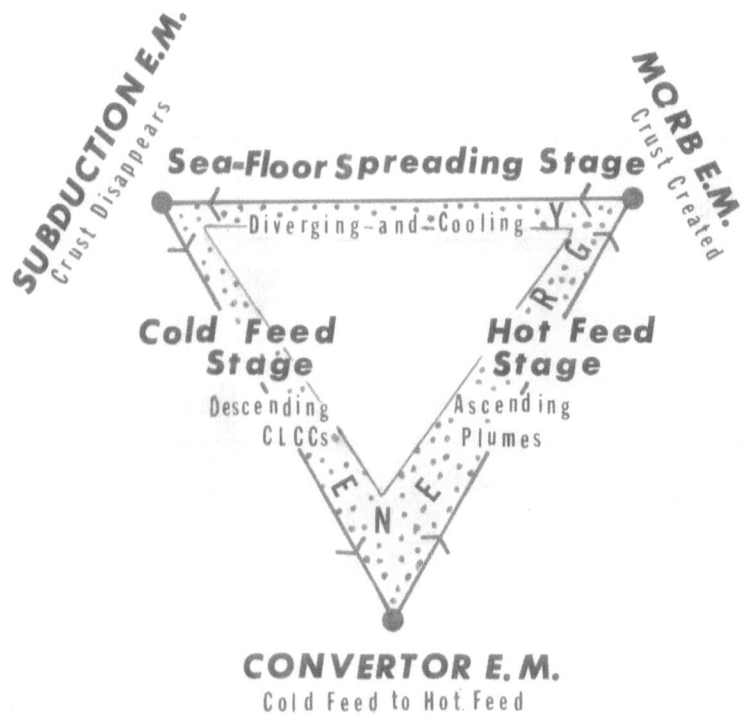

Figure 1 One half of the Earth's plate tectonic cycle (EPTC) is here schematically divided into a three pivotal end member (EM) system with three irreversible connecting stages, note 'one-way' nature of arrows: to first-, the MORB EM; to second-, the Subduction EM; and to third-, the Convertor EM. Material and energy at any point in the system is shown schematically, to be equal to the perpendicular thickness of the outer sides of the dotted triangle. Notice the energy increases at the Convertor EM, since its width of the thickness on the 'Hot Feed Stage' side is wider than that coming down on the 'Cold Feed Stage' side. This represents the conversion process where energy is generated in the Convertor EM and lost throughout most of the rest of the cycle, except for subducted bodies that cold lithospheric crust complexes (CLCCs), do heat up some as they descend into the mantle. If they continue to be colder than the mantle they descend through to the core, otherwise they can stop, at thermal equilibrium, somewhere in between, if there is no other source of density contrast for them to continue on to the core. When the Earth's core is proposed as the Convertor EM it is clear that it must be bounded by reversible reactions, say at the Core-Mantle Boundary and at the Inner-Core Boundary, in order

to manage these two energy extremes and it must also contain an energy source, yet to be discovered. This energy source is to drive the Earth's Plate Tectonic Cycle, which is to convert Cold Feed entering the outer core into Hot Feed that is caused to leave the core, plus an energy source that drives the cores activities for the Earth's magnetic field with all of its variances of strength, focus and polarity. Any other choice for the Convertor EM should likewise be constrained by univariant reactions and it should have an internal source of energy for the conversion of 'Cold Feed ' to 'Hot Feed', plus an energetic means of launching it as an 'Ascending Plume'.

I first draw attention to two relevant observations for the use of experimentally determined stable equilibrium information as regards to the Earth. First, Gibbs, in his introductory remarks, indicates that nature's actions are always toward equilibrium, stable phase equilibrium for example, and that these actions are only opposed by passive resistances that can prevent changes, not just slow them; like the unstable persistence of phases or conditions, e.g., diamonds at one atmospheric pressure and subterranean caverns. When passive resistance is overcome, the action that takes place is typically abrupt and toward reversible equilibrium or toward some other relatively more stable irreversible state, e.g., diamond to graphite and carbon. Carbon forms irreversibly from diamond at one atmosphere pressure on heating as a relatively more stable substance, but graphite forms reversibly from diamond at high temperatures and pressures. On the other hand, sink holes irreversibly fill the unsupported volume of caverns when they tend toward mechanical equilibrium. This tenet means that the cyclic actions of the EPTC, Fig. 1 , can be taken as active tendencies toward equilibrium through its three stages between each of three end members of irreversible activities cycle after cycle ...after cycle toward the logical objective of complete crystallization of the Earth's core. Irreversible heat loss during these cycling processes is the 'times arrow' for the Earth. It is not the degree of crystallization of the Earth's core, for example, because it is likely controlled by a reversible univariant reaction and could decrease or increase in size through time until its complete crystallization occurs. Second, Igneous and metamorphic rock bodies and the earth itself, are testaments to the tenets of Gibbs, as they tend toward equilibrium against passive resistances. This equilibrium is represented by a few essential minerals that are simple to complex in chemical composition. This means that the mineral phases, however complex, can serve as the chemical components of volumes of bivariant equilibria of 'c=p' phases; at their surfaces of univariant equilibria of 'c+l=p' phases; and in their rare three -dimensional lines of invariant equilibrium of 'c+2=p' phases. The Gibbs Phase Rule is

$$F=c-p+2(l),$$

where F is for the variance of the system of phases in mutual contact, 'c' is the number of independent chemical components shared and/or necessary to represent the chemical composition of every physical phase 'p' and '2' is the number of independent thermodynamic variables from the ideal gas law for any homogeneous physical body,

$$PV = nRT \ (2),$$

that typically has three independent variables for a single homogeneous phase or group of such phases. Here 'P'= pressure, and 'V'= volume are two of three independent variables, 'n'= number of moles of the substance, that is fixed if the system is closed, 'R' = the ideal gas constant and 'T'=temperature is the third independent variable. Finally, below is a simple framework of the fundamental bounding chemical components that can be used to portray simple and complex rock and mineral compositions for a direct application of the Gibbs phase rule to the lower mantle and core of the Earth. This chemographic framework will then be used to explore the Earth's current dynamic actions and to follow them during its evolution toward complete crystallization of the Earth's core for a fuller understanding of the evolution of the Earth's plate tectonic cycle.

Early studies of Birch (4) on the physical properties of the Earth reveled an outer core that fits the physical behavior and density of liquid iron except for the fact that its density is 10% lower then anticipated. He followed this discovery with a suggestion that the outer core density is 10% lighter than expected because it is a homogeneous phase alloyed with elements lighter than iron (5). His suggested elements, launched several tens of investigations toward the discovery of such a liquid alloy.

Poirier (6) pulled all of these investigations together for critical review, and concluded that " . . .there is no reason to consider that any single light element is dissolved in the outer core to explain its observed density."; a value confirmed by Mao et al. (7). He suggested that more complex ternary and quaternary systems should be examined by experimentation and by calculation. I agree that more complex systems need to be considered, but I disagree that an alloy of molten iron needs to be found when a heterogeneous mixture of iron-rich liquid and immiscible silicate liquid could explain a 10% decrease in 'bulk density' versus the 'density of a homogeneous phase' and it is unavoidable in relevant iron-silicate systems, Fig.2. The problems are: How does one get silicate liquid into the Earth's outer core? How can it be extracted from the outer core? What role does it play besides lowering the density of the outer core? And how can one recognize it as a product from the core or recognize its effects if and when it escapes from the core? So far, there are two ways of getting silicate liquid into the core, three ways of ejecting it, and at least two means of recognizing it or its related effects. First the most obvious one, if the density of pure iron liquid of the Earth's outer core is 9.8 g cm^{-3}, it can be reduced by 10% by a mixture of 17 vol.% silicate liquid of ~5.7 g cm^{-3} at conditions of the CMB; see (30,34) for density data.

PREVIOUS WORK RELATIVE TO THE EARTH'S MANTLE AND CORE

The first indication that silicate liquid might occur naturally in the Earth's core comes from a density cross-over in the mantle at 7-12, ~10 GPa [215-360, ~290 km according to PREM (8)]. Rigden et al.(9,10) discovered that silicate liquid 'crosses-over from being least dense- to being most dense nonmetallic phase of the lower 90% of the Earth's mantle'. This amazing discovery was confirmed by Agee and Walker (11,12), Miller et al. (13) and Ohtani et al. (14) and means that this dense silicate liquid could be destine for the Earth's core along with molten iron alloy if melting relations, surface tensions, and oxidation states are compatible, and if silicate liquid continues as the densest nonmetallic nonmetallic mantle phase throughout the lower 90% of the mantle to the Earth's core. It should do so since it is apparently the mantle phase with the weakest passive resistance to increasing densification due to its amorphous structure of interconnecting network of tetrahedrons of oxygen about silicon with various cation network modifiers, and likely the first silicate to go to six-fold silicon coordination and higher.

Ringwood (15) opposes such an idea because he concludes that "the Earth's mantle is in disequilibrium with its core" i.e., 'the mantle was never in equilibrium (i.e., had contact) with iron alloy of the core at anytime during its origin because the upper mantle (thought to be typical of the mantle as a whole) is too oxidizing, because it contains magnetite. The upper mantle is also too rich in siderophile elements for the core to have separated from it. These two barriers remain intact today, but the analysis of classical data and more recent data are causes to profoundly amend his professions.

Figure 2 speaks to Ringwood's contention that a magnetite bearing mantle is too oxidized to be related to the origin of the Earth's iron-rich core. He

11

is correct in the subsolidus. Iron and magnetite do not exist in contact, i.e., no Mt + Ir, i.e., no line connects just them, wustite intervenes. But, when the projected bulk composition of the Earth (EBC of Fig. 2A) melts at atmospheric pressure, instead of a simple congruent eutectic liquid formed from positive amounts of each of its subsolidus phases, Ir + Fa +Tr, Bowen and Schairer (16) determined that the contact between Fayalite and Tridymite is replaced by one between Iron and silicate liquid at 'Td'.

Fa + Tr = Ir + L(Fa,Tr,Ir) At T = Td (3)

This silicate liquid 'L(Fa,Tr,Ir) is a 'peritectic liquid' because it is outside of the three phase equilibrium subsolidus phase assemblage of Ir + Fa + Tr which contains the Earth's bulk composition, and in the three phase field of Mt + Fa + Tr because, of its chemical content of ferric iron, under undeniably reducing conditions (16). Having ferric iron in a system with ferrous iron and metallic iron is quite an exception to the general rule that are seen in the subsolidus. One can refer to the difference of this liquid as being 'Mt- normative' verses the Earth's bulk composition that is 'Ir-normative', because other end product solid phases of equilibrium crystallization, Fa+Tr, are common to both. The importance of this peritectic liquid is that, if this liquid is separated from Ir, the Mt -normative liquid would eventually crystallize Magnetite + Fayalite + Tridymite at temperature 'Te', a eutectic, with no phase indication of prior Ir saturation Fig. 2A,2B).

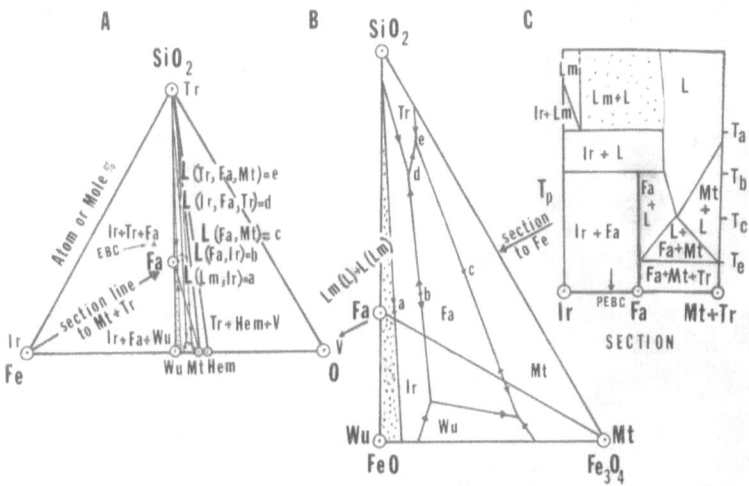

Figure 2 The schematic representation of the stable phase relations for the system Fe-SiO$_2$-O are shown for one bar (specially purified nitrogen) pressure after Bowen and Schairer (16), plus data of Maun (70) on the Magnetite-rich end and Darken and Gurry (71) for the Fe-O binary; except for a large field of two liquids saturated with cristobalite and then tridymite which is of no consequence there. The crystalline phases, their names, their abbreviations, and their chemical formulas are: Tridymite=Tr=SiO$_2$; Iron=Ir=Fe; Wustite=Wu=FeO; Magnetite =Mt=Fe$_3$O$_4$; Hematite=Hem=Fe$_2$O$_3$; and Fayalite=Fa=Fe$_2$SiO$_4$. The shorthand of Greig et al. (72) shows the phases in saturation with phases of variable composition i.e., Silicate Liquid=L and Iron Liquid=Lm. Tb=L(Ir,Fa) represents the temperature, and the peritectic silicate liquid involved with the incongruent melting of fayalite i.e., Fa=Ir+L(Ir,Fa), (Equation 5 in text) and in 2A, 2B and 2C. Tc=L(Fa,Mt) represents the temperature and the point of intersection of the 'temperature composition section' from Fe through Fa, Ta, Tb to 'Tc', a point on the Fa+Mt cotectic in 2A, 2B and 2C. Td=L(Ir,Fa,Tr) is the temperature and the silicate peritectic liquid that represents the upper stability of Fa+Tr in contact relative to L(Ir,Fa,Tr)+Ir i.e., Fa+Tr=Ir+L(Ir,Fa,Tr), (Equation 3 in text) and in 2A and 2B. Te=L(Tr+Fa+Mt) is the temperature and the silicate eutectic liquid representing the upper thermal stability of Tr+Fa+Mt relative to their mutual melting i.e., an eutectic, Tr+Fa+Mt=L(Tr,Fa,Mt), (Equation 4 in text) and in 2A, 2B and 2C. Notice that all silicate liquids lie in the Tr+Fa+Mt triangle or field in spite of the fact that two are in contact equilibrium with Ir and a third is stable with Ir and Lm. EBC is the bulk composition of the Earth projected onto this system in 2A and PEBC is its projection from Tr onto the 'section from Fe through Fa, Ta, Tb to

13

Tc which features the incongruent melting of Fa and the two immiscible liquids above Ta. Only a part of this two liquid field is shown in 2A and 2B by the dotted field. This schematic representation does not take liberty with phase locations, except as noted above, or the sense of temperature (arrows point down temperature) under this isobaric condition but, it is not a point for point translation.

$$Fa + Tr + Mt = L(Fa, Tr, Mt) \text{ At } T = Te \quad (4)$$

Likewise, if Ir separated from silicate liquid, there is no memory of a Fe_2O_3 bearing silicate liquid ever in contact and saturation with it. Notice that this is not an exception but, the rule for this system. Fayalite melts incongruently to a Mt-normative silicate liquid plus Iron at 'Tb' (Fig, 2A,2B, 2C).

$$Fa = Ir + L(Fa, Ir) \text{ At } T = Tb \quad (5)$$

A similar Mt-normative silicate liquid at higher temperature 'Ta', L(Ir,Lm), reacts with Iron to form an iron- rich second liquid,

$$Ir + L(Ir, Lm) = Lm(Ir, L) \text{ At } T = Ta \quad (6)$$

Lm(L,Ir) at Ta (Fig. 2A, 2B, 2C). Above Ta one has [Lm(L) + L(Lm)] where 'Lm(L)' is Ir-normative plus some Fa in equilibrium at one bar pressure, and 'L(Lm)' is Mt- normative yielding lots of Fa, plus Mt and some Tr. If Lm(L) is separated and analyzed at one bar pressure it would have no indication of its former equilibrium with ferric iron bearing liquid. Likewise the analysis 'L(Lm)' would give no chemical or phase signature of its former equilibrium with a liquid Lm. In the absence of high pressure studies on the $Fe-SiO_2-O$ system, one must be alert for the signature of incongruency or of this peritectic process, i.e., ferric iron bearing silicate liquids + iron or iron-rich liquid, in iron plus silicate systems for the application of the crucial incongruities of this system, in ways, as yet, unappreciated since 1932; i.e., Ringwood (16) and many others including myself until just recently. The unusual relationship demonstrated by reactions 3,5 and 6 is;.

$$3Fe^{+2} = Fe^0 + 2Fe^{+3} \quad (7)$$

indicates that some, but not all, of the ferrous iron present converts to iron plus ferric iron.

The Allende CV3, a carbonaceous chondrite meteorite, is an iron plus silicate system worth close scrutiny. Agee et al.(17) studied its sulfur bearing iron-normative composition above its iron sulfide-rich solidus to its silicate solidus and liquidus at pressures up to 30GPa, all at a fO_2 of 'TW+0.5'. At 15 GPa pressure, for example, the iron sulfide solidus is ~900°C below the silicate solidus (1775°C) so, the silicate subsolidus and the hypersolidus relations were studied in fine detail, all in the presence of iron sulfide bearing Iron alloy (FeS,Fe,Ni,etc.) liquid. It is as though there are two separate systems. Indeed, there are two systems of quite limited miscibility, above the silicate liquidus they are composed of two dissimilar parts like [Lm(L) + L(Lm)] of Fig. 2, but here the two liquids are Iron alloy (Fe> Ni> Co) and iron sulfide-normative plus traces of oxygen 0.4 (.2) for Lm(L), and the L(Lm) is silicate-normative and 'probably magnetite-normative' (determination of Fe_2O_3 was not possible) when calculated for equilibrium crystallization at one bar pressure. Interstitial masses and spheroids of quenched iron sulfide bearing alloy liquid are of more than passing interest too because, apparently, they do not settle at all pressures toward gravitational equilibrium as expected by their much greater density than coexisting oxygen-bearing phases!

Shannon and Agee (18) used the sulfur bearing Ir-normative Homestead L5 meteorite, an ordinary chondrite, in a surface tension study between sulfur bearing alloy liquid and its subsolidus mantle minerals at pressures up to 20 GPa (at fO_2 of IW+0.5). They found that the dihedral angle of sulfide bearing alloy liquid with silicates, at 50° to 90° C below the silicate solidus and at upper mantle pressures, was well above the critical angle (108° vs 60°) needed for percolation.

15

This suggested the formation of isolated interstitial patches of sulfide rich alloy melt in a crystalline silicate matrix in spite of its significantly higher density then enclosing solids. As far as I know, Shannon and Agee were the first to predict that " ... melting of silicate minerals of the mantle may be required for efficient planetary core formation. (18)." Presumably, they mean that the melting of silicate minerals at less than 10 GPa would break the interlocking structure of mantle minerals so as to reduce the passive resistance for spheroids to form and fall, coalesce rain drop-like in their course of core fractionation and above ~10GPa in pressure, and below ~290 km in depth, silicate liquid would join sulfide bearing iron alloy liquid in its percolation into the Earth's core . Minark et al.(19) got the same results for surface tension between sulfur bearing alloy liquids and mantle minerals by using a natural 1herzolite, with Fe-Ni-S alloy added.

When the phase equilibria of Fig. 2, and these surface tension results are added to the density cross-over of silicate liquid of Rigden et al. (9,10), for a carbonaceous chondritic Earth (65), it is unavoidable that early formed diffuse domains of dense sulfide bearing iron alloy liquid will flow into the Earth's core through and with a stable dense, and possibly ferric iron bearing silicate liquid, from the lower 90% of the mantle to form the first core of the Earth. For that first core of the Earth of two immiscible liquids, one is likely to contain Fe,Ni,Co,etc.-and FeS possibly K_2S (20) and the other is silicate liqud, and quite likely Mt-normative (see 31 and 34 below and Reaction 7 above), when both are calculated as crystallized at one bar pressure. If this process is a perfect expression of phase- and gravitational equilibrium for phases in contact, the primitive mantle of pyrolite with three parts dunite and one part basalt (for a Iron- and Sulfur free chondrite), plus say 17 vol. % 'basalt' in the core, could yield a first core making-up ~35 vol.% of the Earth, with ~61 vol.% an extractable, immiscible and ferric iron bearing silicate liquid (~ basalt). Because the silicate liquid from the density crossover likely still has a negative buoyancy at CMB depth and beyond, its extraction, toward the current value of 16.2 vol.% core now (21) including 17 vol. % 'basalt' present now, posses quite a problem. My first suggestion is that whole mantle convection following the initial differentiation of the core could yield wholesale interactions with the silicate liquid layer of the outer core that could displace it enough from the core into the lower mantle so that silicate liquid of basic chemistry could react with the lower mantle and be quenched into thermal mantle plumes for ascent (22). Densified silicate liquid can only ascent as a plume if it solidifies, so it must ascent above its stable depth to solidify. Full ascent of these plumes could generate the Earth's first basaltic magma ocean by simple decompression melting (23) in the upper mantle. The first continental nuclei of gabbro-granite are thought to form as a result of igneous and metamorphic processes from the upper 10 % (<290 km) of the primitive mantle that melted and migrated upward while iron- and silicate liquid depleted mantle raises-up beneath this upper most mantle to force interactions with these crustal nuclei and their overlying basaltic crust.

My second suggestion for decreasing silicate liquid of the early core is the heavy bombardment of the Earth-Moon system at 3.9 to 2.8 Ga, according to Cohen et al.(24). It seems reasonable that larger impacts could physically force-up silicate-rich thermal plumes from the outer core by 'sourceward' rebound extrusions from primary pressure effects of impacts on the silicate liquid layer of the upper outer core. These are suggested to yield a third type of basaltic crust like that of maria flood basalts of the Moon. While 'maria' have not been found some very large mafic layered complexes occur in this time frame whose origin is of considerable interest because of their remarkable concentrations of platinum group metals and chromium; the Bushveld in South Africa at 2.0 Byr, the Stillwater in Montana at 2.7Byr, and Russian mafic bodies (with details unknown) and the most probable impact complex at Sudbury Canada at 2.5 Byr whose riches are in nickel and chromium (25). Revised Lu-Hf isotope data of Bizzarro et al. (26) have address the early mantle depletion , place it at 320 Myr after planetary accretion completed in 30 Myr and, after formation of the first solids from the solar nebula at 4,567 Myr. Some time after mantle depletion, the endothermal effects, of convecting depleted mantle and primitive crust into the core to be actively ingested into its body (like Fig. 6), started an inner core like we have today. The inner core is now at only 0.9 vol.% of the Earth, or 5.6 vol.% of the core (21), including half of the transition zone, but only 4.7 vol. % without it. I contend that its formation ushers in what I now recognize as the Earth's plate tectonic cycle (EPTC) with subduction as eventual name for early convection that carries-over from primary differentiation and a continuing means of reintroducing immiscible silicate liquid and other lower mantle phases into the outer core (See Fig. 6) and another internal means of extracting it (See Fig. 7). Estimates of the beginning of the inner core and the start of the EPTC will be discussed below, as will my consideration of its eventual ending at a 100 percent crystalline alloy core (See Fig. 8) and the unavoidable idea that the EPTC was faster early on than it is now, because the primitive core contained a maximum of 61 vol. % silicate liquid and it is unlikely that silicate liquid in the core was perfectly reduced by early convection and meteorite impacts to its current value, even though two stages of basaltic crust may have formed from it. I suggest that there is plenty of silicate liquid in the core to do this since the current amount of the Earth's crust stands at only 1.55 vol. % of the Earth (21), and 61 vol. % in the first core equals ~21 vol. % basalt of the Earth.

Look now at Ringwood's second reason for 'disequilibrium between the mantle and the Earth's core, the 'high concentrations of siderophile elements' in the upper mantle. Li and Agee (27) studied the distribution of Ni and Co between silicate liquid and sulfur bearing iron alloy liquid at pressures of 2 to 20 GPa at 2,000°C, using the Allende CV3 meteorite. Their results, which duplicated the earlier work by Thibault and Walter (28) for a sulfur free system, and the original discovery work by Urakawa (29) at lower pressure, indicate that the expected partitioning of siderophile elements into iron alloy liquids at one bar pressure reverses and goes increasingly into silicate liquids with increasing pressure. This is a very important discovery because if this trend, which starts in the upper mantle, continues to the Earth's core, when silicate liquids are ejected from the core, as proposed below, they would yield a siderophile-rich precursor of basalt in ascending plumes. When these plumes fractionate on reduction of pressure (23) in the upper 10% of the mantle to yield low siderophile content- and low density basalt liquids, they could leave a ponderable enrichment of siderophile elements in what was originally a siderophile depleted upper mantle. So the answer to Ringwood's contention of disequilibrium between the mantle and core of the Earth because of high siderophile elements in the upper mantle, is that he did not account for the actions of siderophile-rich mantle plumes brought to the upper mantle during the EPTC, and especially, he did not anticipate the source of enrichment of siderophile elements in plumes brought from ejections of silicate liquid from the Earth's core, presented herein.

Kesson et al.(30) studied a synthetic MORBasalt composition at pressures of 45, 80 and 100 GPa, using a diamond cell device, in which they encountered the same four phases at subsolidus temperatures ranging form 2,000 to 3,000°K. The four phases, Aluminous Mg,Fe Perovskite, Ca Perovskite, Stishovite and Sodic-Aluminous Ca Ferrite, are the same composition, by energy dispersive micro-analysis, regardless of pressure. In the same pressure range but higher temperature O'Neil and Jeanloz (31) obtained "Iron and glass" for a similar synthetic MORBasalt composition. In a phase equilibrium sense, this would read silicate liquid (=glass) plus Iron or liquid Iron (=Iron) and this is just the type of incongruency that is important for the extension of the phase findings of Fig. 2 to lower mantle and core petrology, namely, the production of Iron and ferric iron from a starting bulk composition that need not contain Iron or Fe_2O_3. This result is like the incongruent melting of Fayalite, $Fe_2 SiO_4$, to Iron plus a magnetite-normative silicate Liquid (Reaction 5 and Fig. 2C) or the reaction of Fayalite plus Tridymite to Iron plus magnetite-normative silicate Liquid (Reaction 3 and Fig. 2B). Both results indicate that for three moles of ferrous iron reduces to a mole of Iron, and two moles of ferric iron, as a general rule (Reaction 7), but all ferrous iron available is not converted to Iron and ferric iron. Here we have evidence for a very unusual and very important relationship for atomic Iron, ferrous iron and ferric iron in these phase equilibrium results, just like those of Fig. 2.

Zerr et al. (32) used a diamond anvil device and physical evidence of melting for a synthetic pyrolite, containing Mg Perovskite, Ca

Perovskite and Magnesiowustite, into the start of lower mantle pressures and then extrapolated their data to the CMB, for a solidus at 4,300°K and 135 GPa pressure, but the exact phase reaction was not given or documented. Their result is in agreement with shock wave data or Holland and Ahrens (33) who gave 4,300°K at 130 GPa for a presumed binary eutectic between Mg Perovskite and Magnesiowustite at the CMB using olivine, $Fo_{90}Fa_{10}$, but the confirming products for this reaction were not recoverable. Kesson et al. (34) got the same three phases as (32) at 70 and 135 GPa using a synthetic pyrolite in their diamond anvil device. But, because they brought their experiments down from above the liquidus into the subsolidus without always allowing enough time for complete crystallization of the liquid present (see glass ? in their Fig. 1), the number of phases and their solid solution compositions may be questioned. Chemical analyses, however, indicate little difference between the three crystalline phase compositions for the two pressures, but another phase was inferred from their analysis of Ni data. They deduced the "presence of a Fe,Ni alloy (see bright specks ? in their Fig. 1) and some compensating ferric iron" (probably in the glass present). Once more there is evidence of the production of ferric iron bearing glass plus iron or iron liquid from a starting material that need not contain neither iron or Fe_2O_3 (Reaction 7). This result is, once more, analogous to the incongruent melting of Fayalite (Reaction 5), and the reaction of Tridymite + Fayalite (Reaction 3) (Fig. 2A, 2B, & 2C). If so, one has evidence for a quaternary reaction at the CMB from their data at 4,300°K and 130-135 GPa (32, 33):

Mg Perovskite + Ca Perovskite + Magnesiowustite = Silicate Liquid + Iron Alloy Liquid (8)

Silicate Liquid is likely a peritectic liquid containing ferric iron and falling on the ferric iron bearing side the plane formed by Mg Perovskite + Ca Perovskite +Magnesiowustite while Iron Alloy Liquid falls on the other iron rich side (See reaction 7 and references 31 and 34). This would be the Iron Alloy absent reaction and if five of these six phases where used for the univariant reaction for the inner core boundary (ICB) there would be an invariant point in the outer core involving these six phases and as many univariant reactions (See Fig. 4). However, it is not possible to proceed at this time without the invariant compositions of the Silicate Liquid and the Iron Alloy Liquid.

The attainment of <u>stable</u> equilibrium is not guaranteed by 'Natures tendency toward equilibrium against passive resistance' noted above. Unstable products from even less stable reactants may be obtained, without contradiction of the 'tendency'. Stable equilibrium products are, however, of upmost importance for deductions and predictions based on experimental data so, for them, there can be no evidence against the supposition that they are repeatable, reversible, and indifferent to starting material, -to viable techniques -to path, -to place, -to observer(s) and -to time. Alas, failure to pass any such test is reason to question the 'true stability' of the particular result. The stability of Mg Perovskite in the above results is based on repeats for three different starting compositions, at several different pressures and temperatures, several different paths, two different places, two different research groups and two different techniques, but none of the results where said to be repeated, reversed or shown to be indifferent to time. This matter comes up because Saxena et al. (35) tested the stability of pure Mg Perovskite at 40 to 100 GPa pressure and temperatures of 1,900 to 3,000° K, using a variety of different starting materials. They indicated that it ($MgSiO_3$) reacted above ~60 GPa and ~2,4000° K to yield post stishovite (SiO_2) plus periclase (MgO),

Mg Perovskite ($MgSIO_3$) = Periclase (MgO) + Post Stishovite SiO_2 (9)

but, they did not test their results by reversing them or testing their indifference to time. Experiments by Serghiou et al. (36) followed on 'oxide powders' heated to 2,800° K at 100 GPa, to 2,600°K at 100 Gpa, and to 2,600°K at 78 GPa and to 2,600°K at 74 GPa for 5, 10 and 15 min., respectively. All results yielded only pure Mg Perovskite and this result is duplicated with other starting materials in the alleged stability field of Post Stishovite plus Periclase of (35). Experiments where Mg Perovskite reacted to this set of oxides

21

(Equation 9) employed techniques similar to those of (35) and they (36) suggested that they resulted from a "disequilibrium dissociation of Mg Perovskite", explained by large temperature gradients due to "unstablized heating techniques". These results mean that experiments with diamond anvil devices are not indifferent to techniques used and that the decomposition products of pure perovskite are most likely at a temperature greater than 4,300°K at the CMB (32, 33),i.e., beyond our particular interest, so there is no evidence against the stability of the Mg Perovskite in the results of (30, 32, 34 and 36 plus reaction 8). However, because it is simpler, and because of its convenience, the hypothetical model I investigate below ignores the presence of both perovskites, having already suggested their future importance for the Earth's univariant reaction at the CMB and their probable involvement in an invariant point in the Earth's outer core.

MODEL PHASE DIAGRAMS AND DISCUSSION

In order to explore the phase relations of the lower Earth where the number of phases is low but chemically complex, a general framework of fundamental chemical components is constructed here so the Gibbs phase rule can be applied to this simple model and to higher order systems of the whole Earth at hypersolidus and subsolidus conditions. Fig. 3 of the M-MO_2-O-M_2O^* system is such a chemographic framework. It includes the neutral to plus one-, to plus two-, to plus three- and to plus four valence states of cations and the neutral to minus one- to minus two valence states of hydrogen, hydroxide, oxygen, sulfur, carbonate anions, other complex anions and gases (37). In detail, it is Fe-SiO_2-O, plus Mg-SiO_2-O, plus Al-SiO_2-O, plus Ca-SiO_2-O, plus Mn-SiO_2-O etc. For easy viewing of this quaternary system as a ternary system, M_2O^* is shown in projection midway on the MO_2-O join. This is justified because alkali oxides have both lithophilic (MO_2) and volatile (O) affinities, and because they are typically low in amount distortion is minor. Notice that Fig. 3 shows the 'metal-normative' bulk composition of the Earth (65), four nonmetal-normative estimates of pyrolite mantle compositions (65) and a 'triangular domain' for the 'nonmetal- normative continuum of fine grained and glassy igneous rocks from basalt, at near the apex, to 'petrogeny's residual system' ($KalSiO_4$ – $NaAlSiO_4$ – SiO_2) of Tuttle and Bowen (38), on the far straight side, plus a host of nonmetal-normative mineral end member compositions for various pressures for general chemographic orientation . Of particular importance here is the wide discontinuity between the continuity of ferric iron bearing igneous rocks, mantle rocks and the iron bearing bulk composition of the Earth. This wide discontinuity across the MO_2-MO join-line has only one means of being connected, and it must be connected. Where else do all of these igneous rocks and mantle compositions of the Earth come from without such a link to its bulk composition? It seems unavoidable to me that the link occurs under hypersolidus conditions where iron and/or iron-rich liquids are in thermodynamic equilibrium and contact with ferric iron bearing liquids. Like those on the Mt-normative side of the SiO_2 -FeO join as shown in the three examples of Fig. 2 (and Reaction 7) and, probably during primary differentiation of the Earth, as indicated above. The 'SiO_2-MO join line is a thermal barrier' for many systems, SiO_2-MgO, SiO_2 -CaO, SiO_2 -MnO systems, across which liquids cannot course by equilibrium crystallization or fractional crystallization, but not the SiO_2-FeO join (see Fig. 2B &2C reactions 3,5,6,7 at one bar pressure) and not for other more complex iron bearing systems synthetic MORBasalt (31) and synthetic pyrolite(34). Both these compositions indicate that the SiO_2-MO join is also not a thermal barrier at CMB conditions for these two crucial compositions that need not contain ferric iron or metallic iron to yield them under experimental conditions (Reaction 7).

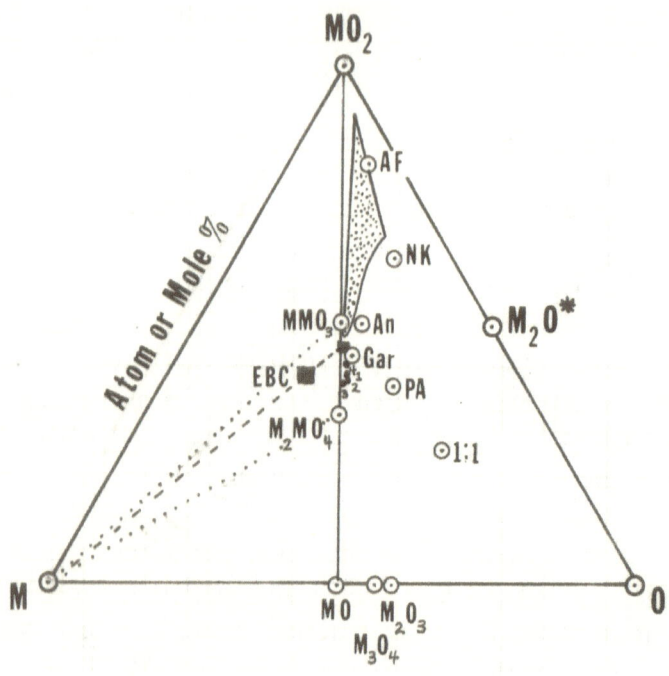

Figure 3 The M-MO$_2$-O-M$_2$O* chemographic tetrahedron with M$_2$O* shown in projection (37) for the use of the Gibbs phase rule where the number of phases is low but of complex chemical composition. It is used here to represent the Earth's bulk composition, EBC (65), four numbered dots for the representation of pyrolite mantle compositions (65) and the 'range of fine grained and glassy igneous rocks of the Earth's crust' from Washington's table of superior analyses (73) after (74) using (37). The location, name, abbreviation, and general formula are given for some important rock-forming mineral end members at low and high pressures: 'M'=native iron, iron-rich meteorites and the Earth's core=Fe, Ni, Co. etc.; 'MO$_2$'= SiO$_2$ TiO$_2$ ect.; 'O'=O^{-2}, S^{-2}, H$_2$O, CO$_2$, etc.; 'M$_2$O'= Na$_2$O, K$_2$O, etc.; 'M$_2$O*'= 'M$_2$O' projected at 50-50 on the MO$_2$-O join (37); 'MO'=wustite=FeO,=periclase=MgO,=magnesiowustite=(Fe,Mg)O, etc.; 'M$_3$O$_4$' =magnetite=Fe$_3$O$_4$, etc.; '1:1'=(Na,K)$_2$O:Al$_2$O$_3$ with SiO$_2$ for 'NK'=nepheline-kalsilite=(Na,K)AiSiO$_4$ and 'AF'=alkali feldspar, albite and orthoclase=(Na,K)AlSi$_3$O$_8$; 'An'=anorthite=CaAl$_2$Si$_2$O$_8$; 'M$_2$MO$_4$'=olivine family and its high pressure forms; 'MMO$_3$'=pyroxene families, ilmenite and high pressure provskites; 'GAR'=M$_3$ M$_2$ M$_3$ O$_{12}$= garnet family and 'PA'=biotite family = K(Mg,Fe)$_3$(Al,Fe)Si$_3$O$_{10}$(OH)$_2$. The several amphibole families of rock forming minerals can be plotted on this diagram too but, are

left off for clarity. The small square on the MO_2-MO join line is the mantle composition if 'M' where subtracted from EBC along the heavy dashed line. The difference between it and mantle compositions one through four indicates that the origin of the Earth's mantle is more complex than simple subtraction of metal from the EBC, and likely includes subtraction of 'basalt' as well. Light dotted lines are possible stability joins with iron.

In the thermodynamic phase analysis of the Earth's major boundaries it soon becomes clear that if one uses the same system of five phases to describe the CMB and the ICB with one phase absent in each, even in this simple model system (where c=3), there must be an invariant point between them inside the outer core of the Earth . This results because, if c+2 phases are used for two surface boundary reactions and one phase is absent at one univariant reaction boundary surface of c+1 phases, say the CMB reaction, and another phase is absent at the other univariant reaction boundary surface of c+1 phases, say the inner core boundary (ICB), it must follow that there is an invariant reaction relationship of c+2 phases, and c+2 univariant reactions, including the two already mentioned, and 'c' other univariant reactions. This is true even though all univariant reactions do not represent physical boundary surfaces and their array in pressure-temperature spaced uniquely describes the chemographic relationship of each phase relative to all others (see reference 75 for details). This unavoidable and interesting set of conclusions are explored in pressure-temperature space, Fig. 4B where c+2 phases at invariance are represented as a 'point' at the intersection of c+2 , univariant reactions, 'as 5 lines', and its c+2 phases are represented as '5 points' on a ternary composition triangle like Fig. 3, in Fig. 4A where 'c'=3.

The chemographic location of three phases of fixed composition (pSt, Mw and Ir) and three phases of variable composition (Ls,Ls* and Lm) in the hypothetical ternary system, Iron-Post Stishovite-Magnesiowustite, are shown in Fig. 4A plus the four phases of the 'Ir and Lm absent' univariant reaction [=(Ir,Lm)see phase reactions listed in Fig.4 with '(phase absent)' notation] which is not part of the invariant point. It is, however, an important univariant reaction suggested here for a lower mantle boundary (LMB) at the top of the D" layer, since an active layer of thermodynamic phases must have two boundaries, of c+1 phases.

The ranking of the seven phases of Fig. 4A from the densest to the least dense at CMB conditions is: iron (Ir)~10.7 g cm^{-3} (39), metallic liquid (Lm)~9.8 g cm^{-3}(39), silicate liquids (Ls & Ls*)~5.7 g cm^{-3} (9

,34, i.e., denser then densest, nonmetal, lower mantle solid phase), magnesiowustite (Mw) =5.68 g cm^{-3} (34), sodic aluminous Ca ferrite (CAF) = 5.48 g cm^{-3} (34) and high pressure post stishovite [name used by (35)] (pSt) = 5.28 g cm^{-3} (34). 'Ir' is located at M in Fig. 4A as a pure metal alloy. 'Lm' is located close to M and is closer to O component than the MO$_2$ component, because the solubility silica and titania in it is likely much lower than that of 'gases' or anions of oxygen, hydrogen, sulfur, carbon, etc. This also means some of the decrease in density of the outer core is due to the alloying 'O and MO$_2$ components' in 'Lm' but no estimate of them is made here except schematically. 'Ls' is located as a peritectic liquid on the O-side of the pSt+Mw join [similar to the location of L(Ir,Fa,Tr)=Td of Fig. 2B] and because the MO$_2$ + MO join is not a thermal barrier for synthetic MORBasalt (31) and -pyrolite (34) at near CMB conditions. 'Ls*' is a eutectic liquid located anywhere inside the Mw + pSt + CAF triangle, here, arbitrarily, closest to CAF in Fig. 4A. Phase 'CAF' is located according to the analysis of (30), using the procedure of (37) as a phase sequestering alkalis in the MORBasalt composition(30).

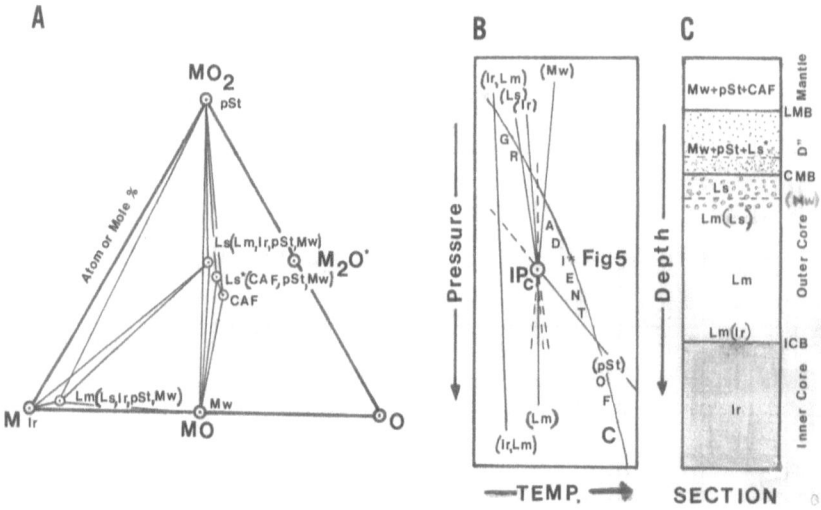

Figure 4 The purposed phase compositions at the hypothetical invariant point IP_C of the outer core are portrayed in the chemographic system of Fig. 3 for the ternary system Iron-Post Stishovite-Magneslowustite part 4A as circled points. The corresponding univariant reactions in pressure-temperature space are illustrated in panel 4B as lines intersecting at IP_C. A hypothetical gradient,'GRADIENT of Core' in 4B intersects five of the six univariant reactions of 4B and these intersections are projected 'straight across' to panel 4C where 'DEPTH' corresponds directly with 'PRESSURE'. They are expressed in a hypothetical 'SECTION' of the lower mantle and core of the Earth as three univariant boundary surfaces and one, the Mw absent reaction '(Mw)' as a labeled dashed line, is used here as a division between the upper- and lower-outer core. Panel 4C is a non-convecting section at perfect gravitational equilibrium and phase equilibrium for phases in contact. This is an important end member for the behavior in this region and one that may be approached but never be attained unless the Earth's magnetic field would cease due to total perfect phase- and gravitational equilibrium as shown, or due to complete crystallization of the core(Fig. 8). From low pressure toward higher pressure, the order of intersection of the univariant lines along the 'GRADIENT of Core' in Fig. 4B is as follows, using the (phase absent) labeling of reactions around the invariant point, after Schreinmakers (75):

(Ir,Lm) Mw+pSt+CAF = Ls*(Mw,pSt,CAF) = Lower-Mantle Boundary =LMB

(Ls) Ir+pSt+Mw = Lm(Ir,pSt,Mw) = No Boundary, unless Ir in the D" Layer

(Ir) pSt+Mw = Ls(pSt,Mw,Lm)+Lm(pSt,Mw,Ls) = Core-Mantle Boundary = CMB

(Mw) pSt+Lm(pSt,Ir,Ls) = Ir+Ls(pSt,Ir,Lm) = Equal to the dividing reaction between Upper- and Lower-Outer Core

(pSt) Ir+Mw+Ls(pSt,Mw,Lm) = Lm(Ir,Mw,Ls) = Inner-Core Boundary =ICB

(Lm) Mw+pSt=Ir+Ls(Mw,pSt,Ir) = Not intersected, No Boundary, but necessary part of the IP_C

Notice with reference to Fig. 4A, that these reactions are of two types: 'eutectic reactions' where three phases surround as a triangle of phases and go into one as in eutectic melting or crystallization i.e., (Ir,Lm)- , (Ls)- and (pSt) reactions or the formation or decomposition of a phase (no examples here). All of the other reactions are 'cross reactions' where the four phases form a four sided figure with the possibility of two diagonals that 'cross' in the center. On one side of the reaction one diagonal pair is stable and on the other side the alternate pair is stable i.e., (Ir), (Mw) and (Lm). All reactions, above, are written from low temperature to high temperature at constant pressure. The first reaction (Ir,Lm) is not part of the invariant point but relative to it , it is both Ir and Lm absent as indicated and thus uniquely labeled. It is the reaction at the top of the D" layer which seems to be required for a layer of thermodynamic phases in a thermodynamically active environment and it is consistent with geophysical results that suggest ~5 km of silicate liquid in the normal D" of 200km(59). The open circles in 4C below the CMB labeled 'Ls' represents 17 vol.% immiscible silicate liquid that can lighten 'bulk density' the outer core by 10% when it is convecting. Ls and Ls* are shown to interface at the CMB in 4C as though they are immiscible liquids. They are not. See their locations in Fig. 4A. They are just illustrated this way to show where they come from, the D" layer and the outer core, respectively. Notice that Lm(Ls) below Ls means Lm is saturated with Ls and above the ICB Lm saturated with Ir is Lm(Ir). In between Lm is bivariate in P-T-X space, see Lm in Fig. 5 for plan view of this section.

It is used for the same purpose here and suggested to be minor in amount since it was not encountered in synthetic 'pyrolite' (34), so a mix of Mw+pSt+CAF is assumed here for the cold lithospheric crust complex (i.e., MORB<<Pyrolite) which gives us a reaction at the lower mantle boundary at the top of the D" layer involving a silicate liquid that others (58, 59) predict to be present there. Phase 'pSt' is located at pure MO_2 in spite of the fact that Kesson et el. (30) found 2.8, 10.3 and 2.7 wt.% $Al_2 O_3$ in it at pressures of 45, 80 and 100 GPa, respectfully.

Gravity is a very important variable in this study because phases Ir, Mw, pSt, Lm, and Ls are basic ingrdients of convection for the generation of the Earth's magnetic field. According to the densities, above, and the phase relations of Fig. 5 there are five mantle seeking phases pSt, Mw, Ls(pSt,Ir), Ls(Ir,Lm) and Ls(Lm,Mw) and one core-seeking phase Ir. Lm(Ir,Ls), Lm(Ls,Mw) and Lm(Ir,Mw) are three different aspects of saturated metallic liquid that defines neutral buoyancy throughout the lower outer core. Above the (Mw) reaction and below the (Ir) reaction of the upper outer core Lm(pSt,Ls), Lm(Mw,Ls), Lm(Ir,Mw) and Lm(Ir,pSt) define its neutral buoyancy and core seeking phase is Ir again but the mantle seeking phases are one less, pSt, Mw, Ls(pSt,Lm) and Ls(Mw, Lm) to make nine phases again.

To evaluate the role of convection, one looks at the effect of each of these nine phases in contact with every other phase below and above the Mw absent reaction. The top half of Table 1 evaluates seventy-two binary contacts below and above the (Mw) reaction during their hypothetical contacts during convection as stable or unstable. Stable contacts are equal to simple physical contacts between phases where there is no physicochemical contact or reaction at the contact other than friction. These are contacts between phases that have solid lines connecting them, twelve different ones are shown in Fig.5, and there are twelve more at temperatures and pressure above (Mw) and below (Ir) for a similar diagram. Unstable contacts are equal to phase contacts between phases that have no solid lines connecting them in Fig. 5, and they are very important because, in the absence of known passive resistances at the phase level at outer core conditions, they yield spontaneous exothermal physicochemical reactions at their contacts as they trend irreversibly toward lower Gibbs free energy , i.e., stable phase equilibrium contacts. This means that any mixture of phases not connected by a solid line at these conditions has higher free energy then those described below them, so free energy is released as heat as stable phase relations are attained. In all, there are twenty-four stable contacts and forty-eight unstable contacts with their 'outcome' suites of convective phase contact reactions toward stable phase equilibrium, to be explicit, or just phase contact reactions (PCRs) for short. The outcomes of these reactions vary in kind depending on the phase field intersected by the bulk composition of the contact pair, for example combination #8 in Table 1 between pSt and Mw yields five different phase assemblage outcomes whereas #40 between pSt and Mw only yields three different phase assemblage outcomes, three of the five for #8. The relative size of the contacting phases yields an instantaneous bulk composition for the contacting pair and governs the phase make-up of the stable equilibrium outcomes according to the prevailing phase diagram at the actual pressure and temperature of the contact. The extent of destruction of the phases involved and their total mass governs the amount of heat evolved in the PCRs, but this aspect is not dealt with here is this hypothetical system. See the legend for systematic mixtures of Mw and pSt along the dotted line in Fig. 5.

Immersion of crystals in Ls and Lm means that these two liquids entirely fill the space of the outer core, so there are no strictly binary collisions between crystals where Ls and/or Lm are not included too.

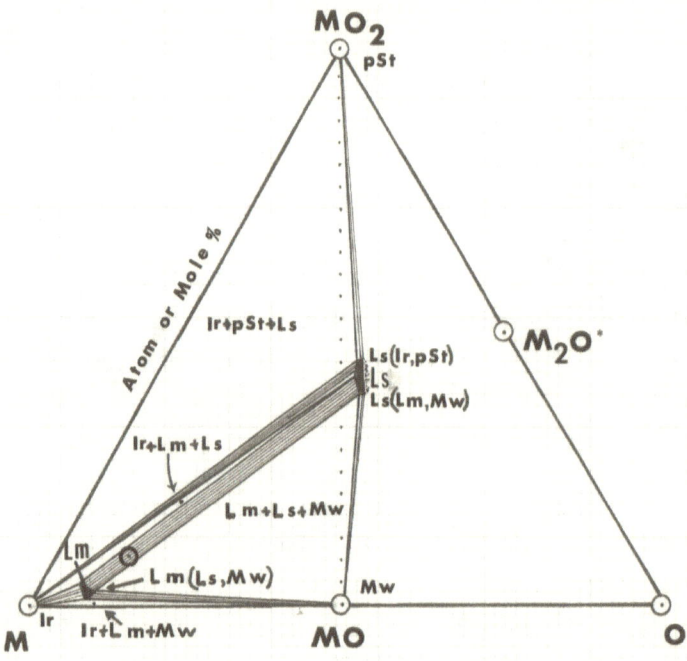

Figure 5 The general composition tetrahedron is shown as a triangle in projection as done in Fig. 3 to show the stable ternary phase relations for the hypotheritical system Iron-Post Stishovite-Magnesiowustite under isothermal and isobaric conditions in the Earth's outer core at the ' * 'on the 'GRADIENT of Core' in Fig. 4B. All three phases of fixed composition, pSt, Mw and Ir are stable throughout the outer core as are the contacts between pSt+Ir and Ir+Mw but the contact between Mw and pSt is not stable. There is no solid line connection them as there is for the other two pairs. The dotted line connecting Mw and pSt is for the purpose of illustrating the changing bulk composition between these two ends (A and B Table 1, combinations #8, and #40--- for phases pSt and Mw) and the intersection of stability fields along its course of bulk composition change are discussed below. During convection, when contacts of phases are considered with every other phase, it is not just 'Ls' that is considered but its isothermal isobaric invariant compositions at various temperatures and pressures, i.e., Ls(pSt,Ir), Ls(Ir,Lm) and Ls(Lm,Mw) for the * along the gradient in the lower-outer core, i.e., below (Mw) and above (pSt). According to the shorthand of Greig et al. (72), this reads, "Ls saturated (equals, in thermodynamic contact) with pSt and Ir, Ls saturated with Ir and Lm and Ls saturated with Lm and Mw" i.e., each of the three corners

of the Ls saturation surface in the figure. Likewise, it is not Lm alone that needs to be considered during convection contacts but Lm(Ir,Ls), Lm(Ls,Mw) and Lm(Mw,Ir) at the three corners bounding the bivariant region of Lm. This means that there are thirty-six binary contacts among these nine phases in Table 1, of which, only twelve are stable contacts, meaning, connected in the phase diagram with unique solid lines. Actually, there are two other solid lines, making fourteen solid lines in all for the diagram. Each of these 'other thinner lines' is on the 'O' side of the diagram. They are thinner two phase lines, one set for pSt+Ls(pSt) and one set for Mw+Ls(Mw), which are used to finish off the diagram on that side and both are part of what are called 'ruled surfaces lines' which are like the univariant region Lm(Ls)+Ls(Lm) where the 'rulings of the surface between the two phases in equilibrium' shows better, plus others at Mw+Lm(Mw), Ir+Lm(Ir) and Ir+Ls(Ir). For the remaining twenty-four unstable contacts, they can be said to react spontaneously, irreversibly, and exothermally toward stable equilibrium, and lower Gibbs free energy, according to the bulk composition of the reacting pair. Taking all mixtures of Mw and pSt along the dotted line, for example, when a small crystal of Mw encounters a large crystal of pSt , they react to yield pSt+Ls(pSt,Ir)+Ir, as Mw and part of pSt are destroyed by the formation of Ls(pSt,Ir) and Ir at their former contact. The next change occurs when the two crystals are closer in size when both are destroyed to yield Ls(Ir)+Ir and then again when Ls(Ir,Lm)+Ir+Lm(Ir,Ls) form; then again when Ls(Lm)+Lm(Ls) form. Beyond this relatively 'broad two phase' outcome, larger crystals of Mw react with small crystals of pSt to yield Ls(Mw,Lm)+Mw+Lm(Mw,Ls) when pSt is destroyed and only part of Mw is destroyed as the contact reacts and the bulk composition shifts in favor of pure Mw and the completion the five 'outcome suites' for combination # 8 of Table 1; all in the lower part of the Earth's outer core. In addition to the thirty-six contacts mentioned above between the (Mw) reaction and the (pSt) reaction (#1 through #36 in Table 1), there are thirty-six additional contacts between the (Ir) and (Mw) reactions in the 'upper outer core' of which twelve are stable and twenty-four are unstable (Table 1). These results may vary some depending on the details of the reference isothermal-isobaric section drawn but the results of combination #40 are straight forward and yield only three reaction outcomes, all three of which also occurred as part off the 'outcome suites' of phases in combination #8, the first , fourth and fifth.

The fact that 83 vol. % iron liquid and 17 vol. % silicate liquid fill the entire outer core suggests that all contacts are immersed, thence buffered by the mutual presents of these liquids (Ls and/or Lm). For example, combination #94 is the Ls-buffered equivalent of combination #8 and #75 is the buffered equivalent of combination #40. The only difference is that one imagines that the dotted line is not exactly between pSt and Mw but between a point on the pSt+Ls line near pSt to a point on the Mw+Ls line near Mw which is represented as [pSt+Ls(pSt)]+[Mw+Ls(Mw)] = pSt+Ls(pSt,Ir)+Ir as Mw and some pSt are destroyed; Ls(Ir)+Ir as both Mw and pSt are destroyed; Ls(Ir,Lm)+Lm(Ls,Ir)+Ir as both Mw and pSt are destroyed; Ls(Lm)+Lm(Ls) as both Mw than pSt are destroyed; Ls(Mw,Lm)+Mw+Lm(Mw,Ls) as pSt and only some Mw are destroyed; and Mw+Ls(Mw) as the hypothetical point on the Mw+Ls join line is attained at zero pSt+Ls(pSt) . The difference is that the end points are represented as two, two phases mixtures inside of the 'brackets' as [pSt+Ls(pSt)]+[Mw+Ls(Mw)] to indicate immersion buffering, i.e.,contact, of both phases by Ls in this case. Contact #94 could happen anywhere between reactions (Mw) and (pSt) just like combination #8 but being buffered is more realistic of a 'liquid dominated regime' of the outer core. Combination #94 is also particularly important because it starts with two buffered mantle-seeking phases, that yield mantle-seeking plus a core-seeking phase Ir or a dense phase like Lm that would both tend to separate from a Ls rich domain that tends to create local convection or supplement existing convection plus there is the generation of heat as these new phases are created as an expression of their loss of Gibbs free energy as stable phase equilibrium is attained, at least temporarily. The only binary phase contact reactions of the upper part of table 1(#1 to #72) that remain relevant are the twenty-eight unstable liquid-liquid combinations.

Note the circled dot toward the bottom of the Ls(Lm) + Lm(Ls) 'ruled surface'.It schematically represents the bulk composition of the Earth's outer core in this hypothetical ternary model system and suggests to me that one has two mega-bodies of Ls at <17 vol. % and Lm at < 83 vol. % at centimeters to kilometers in size. These bodies convect and interact with one another as they each contain 'micro-phases' of Ir, pSt, Mw and Ls (in Lm) and Ir, pSt, Mw and

Lm (in Ls) at micrometers to millimeters in size. These micro-phases which are present in significant to copious quantities in each or the mega-bodies and interact with one another as friction for twenty four stable contacts, phase contact reactions toward stable equilibrium for twenty-eight unstable Liquid -Liquid contacts and thirty-seven unstable liquid -buffered contacts, plus 'gatherings and splittings'(see text) during convulsions and differential movements of these two bodies.

Further analysis of liquid buffered ternary contacts is quite complex. Contemplation of the simple contacts of pSt+Mw+Ir, for example, covers every composition in the entire ternary if one would consider every possible combination, as done in Table 1 for every binary liquid - liquid contact and every liquid buffered binary contact. Needless to say, there are certainly liquid buffered ternary collisions and they too contribute as exothermal heat sources of the outer core.

Notice that there is an overall equilibrium product of Lm(Ls)+Ls(Lm) for the circled bulk composition in Fig. 5. My interpretation of this, as an equilibrium tendency, is that there are immiscible connections on two scales: the 'mega-scale' ,called mega-bodies of Ls at < 17 vol. % and -of Lm at <83 vol. % because, they are not homogeneous phases throughout each mega-body. This is so because they contain inclusions of micro-phases of crystals and reciprocal immiscible liquids at the micro-phase scale. The 'micro-phase scale' is the other scale, and thought to be from micrometers to millimeters, whereas Lm- and Ls mega bodies are thought to range in size from centimeters to kilometers with tendencies 'to collect or to gather together' on contact with themselves and to 'split and divide' each other along differing contacts. Both liquids 'bodies' are thought to contain significant to copious quantities of crystals of pSt, Mw and Ir plus small sizes of Lm in Ls and of Ls in Lm as they are produced at the 'micro-phase level'by PCRs during convection. Lm mega-bodies are thought to convect from the hotter inner core to the CMB where they are translated along it conducting heat toward a cooler mantle then descending as a cooler mass toward the inner core. There is a natural tendency for both liquids is to carry as many micro-phases as possible. Ls mega-bodies are a special case because they have a definite 'mantle seeking tendency' that is increased by the presents of pSt and Mw and decreased by Ir. It has at a phase density contrast of about two with Lm bodies, and it is imagined that it tends to migrate in larger sizes to the CMB, to 'wet it' and to stay there as 'smears' that gather other 'free-floating spheres' of different sizes which explode into one another when joined. This is an exothermal event too since the micro-phases in each of the combining parts that are brought into sharp contact during the 'gathering process' or during the 'splitting process' for that matter. So, once there, at the CMB Ls mega-bodies tend to stay there, and 'wet the underside of the CMB', rather than just touching it as in the high surface tension case of Lm mega-bodies, whose contacts with Ls-bodies of the CMB tend to split them, squash them and divide them generating heat by micro-phase contact reactions toward equilibrium. The CMB is believed to consist of interlocking pSt and Mw in the model system but its actual thickness is unknown. It must be thick enough to keep Ls* from the lower D" layer from freely mingling with Ls of the outer core and vice versa. Reactions at the micro-phase level throughout each of the mega-liquid-bodies might resemble a 'mid-western meadow on the night of a new Moon in mid-summer with lightening bugs going on and off, on and off as far as the eye can see--- with frictional contacts, gathering contacts and splitting contacts going on and off and, on and off in a more intense fashion.'

The other change that effects the amount of crystals in the mega-bodies is the changing size of the saturation fields of Ls and Lm. They 'broadening with depth' and 'narrowing toward the CMB' as their temperatures approach 'boundary conditions'. Broadening results in dissolving of included crystals, an endothermal process, and narrowing results in crystallization as saturation fields narrow, an exothermal process. This complicates the picture a bit, but it is a necessary complication. This natural dynamic is a major part of what restricts the composition of the ejected silicate liquid that shows up in the restricted nature of FOZO- and C-component discussed below as the signatures of silicate liquid ejected from the Earth's outer core.

On the other hand, what happens when convection slows or stops? At perfect gravitational- and -phase equilibrium for phases in contact, a vertical phase configuration like that of Fig. 4C results with crystals of pSt closest to the CMB, Mw below and Ls below them and although this is a very unlikely outcome it is important as an end member in the behavior of all phases given enough time and quietude to equilibrate. It also suggests that as convection lessens and the Earth's magnetic field lessens, Ls-bodies have a great propensity to stagnate and collect at the CMB, where convection at the micro-phase level becomes important. In brief, assuming all phases are above the true Brownian motion size, Ir settles toward the core through Ls- and Lm bodies with likely encounters with pSt and Mw as they ascend through Lm- and Ls bodies toward the CMB. Note however, this pair is not stable in contact, with pSt overlying Mw below theCMB. Their reaction through combination #79 of Table 1 has three outcomes, each producing heat, which increases the temperature in the vicinity of the CMB along with more Ls and Lm plus either pSt or Mw. But, the roof of the CMB is itself composed of pSt + Mw on the cool side of the Ir absent reaction (Fig. 4). If enough heat causes pSt and Mw to develop Ls and Lm, they develop between the crystals and that can certainly weaken the CMB. It 'weakens the boundary' in whatever shape the Ls mega-body may take. They could be in local concavities or along curved concavities that reflect pressure build-ups in the outer core prior to failure at such points- or lines of weakness of the CMB.

Finally, for what may be Natures quaternary system, suggested above in reference to the likely reaction for the CMB, there will be 135 binary contacts in each of two or three parts for the outer core, of which one third would be stable yielding contacts for frictional heat and the remaining unstable binary contacts are suitable for analysis for liquid-liquid reactions and liquid buffered phase contact reactions with hundreds of different outcomes, each heat producing to varying degrees throughout the entire outer core. Radioactivity of Ls bodies , other mantle phases and perhaps still Lm (20) is limited by a decay rate whereas phase contact reactions and frictions are not; they do not 'die, tire or wear-out'.

Of course, the energy produced by liquid - liquid PCRs, liquid buffered PCRs and friction vary with the total concentration of 'convectables' and that could yield a spectrum of energy values for the entire core from subnormal - to normal - to an aberrant high energy state. Subnormal energy is imagined to be where ingestion of cold lithospheric crust complexes (CLCCs) 'swamps the energy' of the outer core by its endothermal ingestion processes, Fig. 6. This generates a tendency toward closure of the outer core by favoring crystallization to the inner core via the (pSt) reaction. However, note as Ir forms onto the inner core, Ls(Ir,Mw) and Mw also form at the ICB as mantle-seeking micro-phases in Lm mega-bodies. This is a realistic phase action rather than the 'pseudo-phase action' of a Ir depleted metallic liquid known as 'chemical convection' (43). Because convectables can become very high under the circumstance of subnormal energy for the core, the outer core energy is liable to swing past normal to strongly aberrant energy by the action of numerous convective PCRs. If one defines normal energy for the core, as a heat content low enough to crystallize Ir onto the inner core and still absorb and conduct its heat of crystallization away in its mantle seeking phases, in Lm mega-bodies, then the core and lower mantle boundaries change very slowly toward closure of the outer core. An aberrant high energy state for the entire core is one where normal boundary reactions are reversed as the outer core grows at the expense of the inner core and mantle (See Fig. 7).

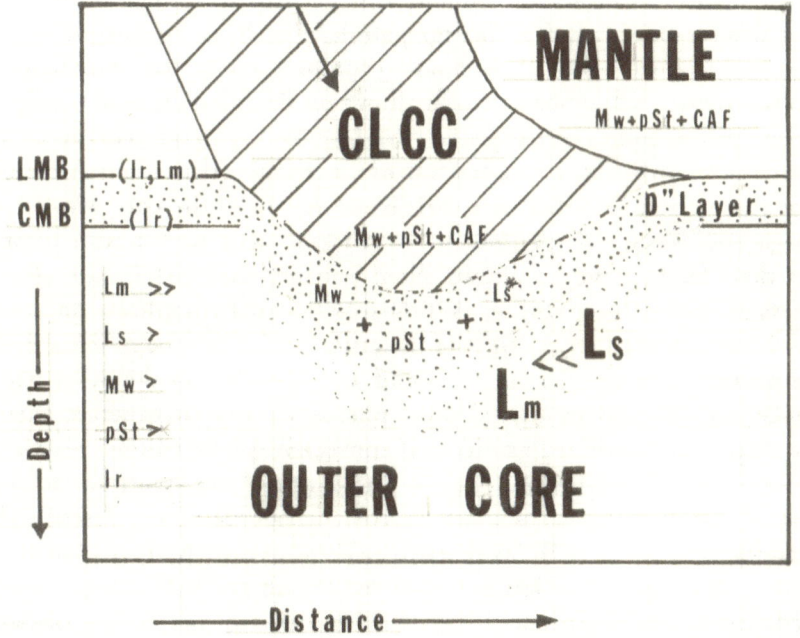

Figure 6 This diagram is a schematic 'stop action view' of the 'ingestion facies' of the D" layer of the lower mantle where a subducted cold lithospheric crust complex, CLCC, has broken into the Earth's outer core and is braked and turned by it and baked by it at its contact. It breaks into the outer core because of the inertial energy of the complex is thought to carry it through the CMB then it is slowed and turned by denser Lm of the outer core and it floats on Lm as it is heated endothermally by the outer core. Ls*, Ls, Lm, Mw, pSt and Ir are 'formed or freed by reactions (Ir,Lm) of the LMB and (Ir) of the CMB plus the (Mw) reaction dividing the upper outer core from the lower outer core. For example, phases pSt and Mw are first released as they melt at contacts with phase CAF yielding Ls*(CAF,Mw,pSt) among them and consuming CAF in the process to leave Ls*(pSt,Mw) + pSt + Mw. At sightly higher temperatures and pressures pSt+Mw contacts become unstable and their contacts are destroyed when replaced by Ls(Lm)+ Lm(Ls) due to the (Ir) 'cross reaction'. This reaction is normally associated with the CMB (see (Ir) Fig. 4) and leaves Ls(pSt) + pSt and Ls(Mw) + Mw or Ls(Lm)+Lm(Ls) as pSt and Mw are set free from one another. So the freeing of lower mantle phases during the ingestion of complexes in the outer core of the Earth is unavoidable and very important because all the crystalline phases of the

lower mantle and core (Ir, pSt. and Mw) are stable throughout the core. For this reason, I call this 'disaggregation by melting'. In the vicinity of the 'keel-shaped region' of the floating complex, the typical abundance of included immiscible phases and Ir shown on the left margin of the figure for mega-Lm bodies is changed to yield Ls-rich mega-bodies because Ls is much closer to the bulk composition of the complex than Lm is. Although not shown here in detail here, this is the region of the core where Lm-mega-bodies can strongly interact with forming Ls-mega bodies to pick up their loads of micro-phases. Note however that CLCCs are of two parts MORBasalt<<pyrolite, as 'lithospheric-crust complexes' and both are likely attached by the heat of the outer core. Here the difference can be portrayed as pSt-rich on one hand and Mw-rich on the other hand. Once released and immersed these phases are free to generate energy by their liquid-buffered exothermal convective 'phase contact reactions' toward stable equilibrium. The size and shape of the ingestion facies of the D" layer is designed as a fore-shortened example of that described by Kendall and Silver (42), 'as sheet-like, being several hundred km thick and up to thousands of km in length and breath and indicating evidence of liquid internally, probably like Ls*. If they are correct, the size and shape of the 'complex' with time should increase or remain the same, i. e., as a graveyard on the Earth's outer core!. If the ingestion scheme above is correct, the expression of the complex eventually disappears as it is consumed by disaggregation melting into the outer core and , a new D" layer is formed of it. And, the Earth's plate tectonic cycle continues rather than being 'hung-up' for lack of sustenance.

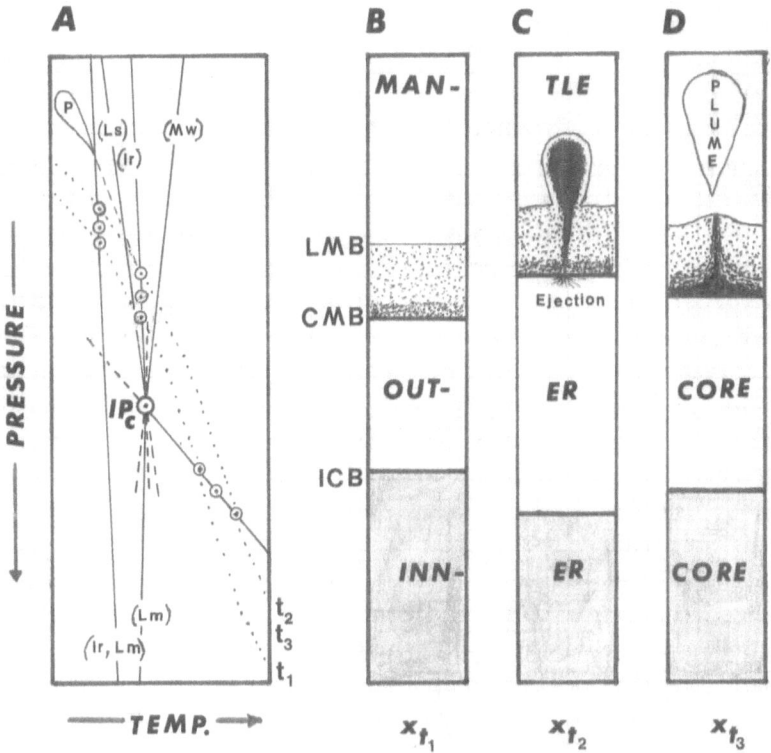

Figure 7 This figure 7A illustrates a pressure-temperature diagram like Fig. 4B for 'the hypothereotical ternary system Iron-post Stishovite-Magnesiowustite' that has a pressure scale equal to the depth scales of 7B, 7C and 7D (for simple horizontal transfer of intersections in 7A to boundary locations in 7B, C, & D) and they have a short arbitrary distance 'x' as their horizontal axes. Two gradients 'at times t_1 and t_2, are shown in 7A, 'the third at time t_3' is midway between the other two. Fig. 7B displays a 'normal facies' for the D" layer with a minor amount of silicate liquid (Ls* Fig. 4A, 4C) (solid black to black stippling) in the lower 5 km (~2.5 Vol. %) of a typical 200 km thickness worldwide (59) which I use to justify the (Ir,Lm) eutectic reaction. Between Figs. 7B and 7C it is imagined that the heats of radioactivity, spontaneous exothermal heats of Ls and/or Lm-buffered 'phases contact reactions'-, Liquid-Liquid 'phases contact reactions' toward stable equilibrium, and fiction between stable phases accumulate until gradient t_2 would be attained, if, and only if, the mantle follows the tendency of the outer core to expand ideally to this new stable dimension. Instead, it is suggested that the mantle resists

this tendency to expand due to its passive strength to remain unchanged. Over-pressure develops in the outer core due to the +5.0 cm^3g^{-1}, of iron melting alone (76), as the (pSt) reaction of the inner core boundary reverses from crystallization to melting and it is another cause of this expansion tendency of the outer core. Expansion pressure continues until the mantle's limit of strength is exceeded and it fails abruptly at a point of weakness or along a line of weakness. Failure yields a sudden decrease in pressure of the Earth's outer core as Ls is driven out of the outer core with varying degrees 'explosiveness' and, it encounters the lower mantle with the vigor of a superheated fluid under pressure to attack grain boundaries at three and two phase junctions (black = Ls attacking mantle in Panel 7C). It eventually exhausts its excess heat to become a hybrid, hot and solid domain (open'tear-drop'labeled 'PLUME' in Panel 7D). The domain is hybrid because it is formed partly from the core and partly from an indigenous lower mantle, and generally rich in siderophile elements (27,28,29). Where lines of weakness were involved, major transfers of material and energy are thought to become 'curtains-like regions' of ascending plumes. Curtains of plumes continue to ascend eventually melting on decompression (23) to fractionate siderophile elements into the upper mantle as low siderophile magmas service segments of the Mid-Oceanic Rise, the MORB End Member of the EPTC(Fig. 1). Explosions at single points of weakness of the CMB behave in a similar fashion on a larger- or on a smaller scale. They are single plumes that add low siderophile basalt magma that are added episodically as floods of basalt or as lesser amounts to basaltic oceanic island 'hot spots'. Hot spots, that Williams et al (57) link with 99 % certainly to 'ultra low velocity zones' (ULVZ)s below them in the D" layer, where the amount of liquid in its 200 km expands from 5 km and 2.5 % (59) Fig. 7B, to 45 km(58) Fig. 7D or about 22.5 % of the total D" layer. Figure 7D along the intermediate gradient of 7A is taken as a 'relaxed stage' of the D" layer where it is suggested that the ejected silicate liquid, not included in the plume, settles back toward its exit point to create an accumulation of Ls*. I suggest that this silicate liquid-rich stage, this 'relaxed stage' is the 'ejection facies of the D" layer' (because the actual ejection facies is too short -lived and likely undistinguishable) that is the 'ULVZs' described by geophysicists (58). Further it seems obvious that it represents a 'wound' from the active ejection of Ls rather then a source of a thermal plume itself. The idea that plumes could form from ULVZs is lacking in impetuous to form a viable plume, i. e., a dense silicate liquid at CMB depth must solidify to ascend and ascend to solidify. Furthermore, I predict that 'elongated wounds' in the D" layer beneath the Earth's ridge and rises system will eventually

be found since I contend that the basaltic crust for them originates in the core and lower mantle beneath them, and that their location is likely controlled by expansive features on the Earth's CMB i.e., as templates for the EPTC rather than any other top-down structure of the Earth's crust and mantle above them.

Ejection of Ls from the Earth's outer core is the 'Hot Feed Stage' of Fig. 1 is schematically illustrated in Fig. 7 using three steps that are three new facies of the D" layer. They are the normal-, the ejection- and the post-ejection facies. The actual ejection facies shown is thought to be too short a time to be recorded geophysically. Ejection of Ls from the outer core takes place when the thermal state of the outer core is significantly higher than normal. Such an aberrant energy state is imagined to ensue due to a higher than normal concentration of convectables of all types, from an independent rise in ingestion of CLCCs. The unavoidable liquid buffered PCRs and liquid-liquid PCRs produce so much heat, or produce heat so quickly inside and along mega-bodies, that the outer core tends to expand due to reversals of the boundary reactions; notably the +5.0 cc/g (76) from iron melting alone as the ICB reaction. This results in an expansion of the outer core against the passively resisting strength of the Earth's mantle. When spontaneous failure of the passive resistance of the CMB occurs, Ls is ejected more of less explosively through a 'point or line of weakness on the CMB' into the lower mantle (Fig. 7C). There it is imagined to flood into the lower mantle under pressure as a superheated flux with a high appetite to be quenched by its endothermal activity of dissolving lower mantle phases at and along grain boundaries. Ultimately it forms a hybrid, hot and solid domain of basic chemistry rich in siderophile elements. Such a domain is imagined to ascent as a hot and low phase density basic plume in a denser and colder ultrabasic mantle. In the shallow upper mantle one imagines melting due to decompression (23) to yield basalt liquid plus a siderophile enriched residue in the upper mantle during fractionation. These basaltic liquids are thought to intrude as 'sheets among the sheets of sheet complexes' or as 'pillows among pillows of pillow complexes' as 'newly created oceanic crusts' at the MORB end member of the EPTC (Fig. 1) from 'curtain-like plumes' from line defects of the CMB. Lessor 'point derived plumes' from the CMB ascend after similar processes in the lower and upper mantle. They are exemplified by typical episodic volcanic actions at oceanic island basalt 'hot spots' and even some are of a continuous nature like Hawaii's Kilauea's Pu 'u' O'o crater's eruption of over two decades now and the historically significant Iceland volcanic complexes. Flood basalts from larger point defects are also possible but supporting geochemical signatures connecting them to the Earth's outer core are not established at this time.

The concentration of silicate liquid above the CMB in Fig. 7D, the (post) ejection facies, is taken to represent a low melting dense silicate liquid not included in the plume which is Ls*-rich from the (Ir,Lm) reaction that peaks with the ejected Ls into the lower mantle. As a lower melting liquid (Fig. 4) it is present within the pSt + Mw of the lower periphery of the hybrid Ls which solidifies, but Ls* remains liquid on the bottom side and settles back to the region of the weakness on the CMB. Such a liquid-rich zone in the D" layer would be referred to as an 'ultra low velocity zone' (ULVZ) by geophysicists (58) and should be thought of as 'a wound of silicate liquid from the ejection of silicate liquid from the outer core' rather than the source of the resulting plumes. Williams et al. (57) have established a quite significant positive correlation between oceanic island basalt 'hot spots' with ULVZs below them, but no ULVZs have been discovered as yet below the global ridge and rise system in the D" layer. Courtillot et al. (59) only found nine 'hot spot' plumes that had continuous roots to the Earth's core, thirteen were based in the upper mantle and twenty-five others were based in the lithosphere. In Montelli et al. (60) only eight oceanic island basalt 'Hot Spots' plumes were based at depths as great as 2800 km while twenty were 1000 km or deeper and four were less than 1000 km. The biggest disagreement in these two studies was the inclusion of Iceland and Hawaii as deep seated plumes in (59), but (60) indicated that, "... the strong velocity anomaly observed in the upper mantle beneath Iceland is not generated by a large upwelling from the lower mantle." And, I might add, at the present time, because of the basic nature of oceanic island basalts 'Hot Spots' is to be episodic and Iceland, with its long history of volcanic activity could be a good case of this. Even Hawaii, that has been erupting continuously for twenty years, could only be carried down to >2350 km by (60), until the area was extended to a wider region at 2800km. Tests along the Atlantic ridge and rise system at 15°N and 25°N and along the Indian Ocean ridge and rise at 35°S all had plumes to 1900km or deeper but, none of them had ultra-low velocity plume regions in their lower 1,000 km.

Some insight may be gained by sharing a petrologist idea of how so-called 'ridge push' might result during the intrusion of 'sheets of basalt among sheets of basalt' along the mid-oceanic ridge and rise when 'slab pull' is not present. First I would say that the general intrusion of basalt transforms the ridge and rise to a simple rise (e.g., The East Pacific Rise) via the expansiveness of the intrusion. Tension cracks along the crest can be thought to 'grow downward' till they are tapped and fill by the intrusion of basalt. In detail, as the rise forms it seems clear that points just off the axis move 'up and out'. Soon after intrusion it is supposed that an unsupported volume develops in the deep part of the rise as liquid has migrated up at the rise and up and out beneath the rise along the asthenosphere. Paired axial faults along this 'rise' causes it to become a 'ridge' again as a 'keystone structure' constrains further substance to go 'down and out' instead of its reverse path of 'down and in'. Instead the 'out' component of the original expansion movement is expressed during outward spreading as the flanks subside toward gravitational and thermal equilibrium. The process resembles that of the 'long shore drift of sand' along our coasts when the wind is not alined with the offshore dip of the bottom.

If the strength of Earth's magnetic field is effectively decreased by the forceful ejections of liquid silicate from the core, it is suggested that this may yield a 'window of time ~7,000 yrs (40)' where the polarity of the Earth's magnetic field could change as the strength of the field reorganizes while recovering from these ejections. The abrupt ejection of silicate liquid relative to the Earth's axis of rotation could produce changes in the length of day that have been noticed in the Earth's rotation or geomagnetic jerks such as occurred in 1999 (41). Thermal plumes may be added to the effects of slab subduction which are already related to changes or trends in polar motion as redistributions of Earth mass (41).

Because the inner core plays such an essential role here in the ejection of silicate liquid from the Earth's core, when did it and the EPTC start? And when and how will it end? Fig. 8 illustrates how the crystallization of the core could end at 100% crystalline Ir core in contact with Mw + pSt + CAF at a liquid absent low free energy relationship. This would be the trend of the core in the absent of recharging CLCCs from the subduction end member of the EPTC or it could truly become the 'graveyard' for subducted slabs (42) until they stop coming.. Buffet et al.(43) calculated the start of the inner core using first principles. They predicted that the Earth's first core started at 2.8 Ga. but, neglected to indicate when it would finish. If one assumes that the inner core occupies 5.6 vol.% of the core now, including half of the transition zone (21), the same linear rate to completion yields 50 Ga . Layer et al. (44) have evidence for what may be the earliest magnetic reversal at 3.2 Ga and there are those who believe that a central core is necessary for a stable magnetic field that does not reverse too frequently (45). In this case the same rate to complete core crystallization finishes at 57 Ga. Zircons dated at 4.3 and 4.1 Ga 'from a chemically evolved terrain', according to Bowring and Housh (46), yielded 78 and 73 Ga for their simple completion ages. This strongly suggests that the EPTC is close to being a perpetual motion machine producing prodigious amounts of heat, doing vast amounts of work, creating, recycling and recreating . . . and recreating oceanic crust and lithosphere and its metamorphic equivalents of cold lithospheric crust complexes with only a minor amount of inner core crystallization. One variation on this proposed constant rate has to do with the concentration of immiscible silicate liquid in the outer core, however depleted it may be after early whole mantle convection and meteor bombardment (24) acted on the original amount of up to 61 vol. % liquid average over time of the crustal sources that contribute silicate liquid to the outer core during ingestion after subduction.

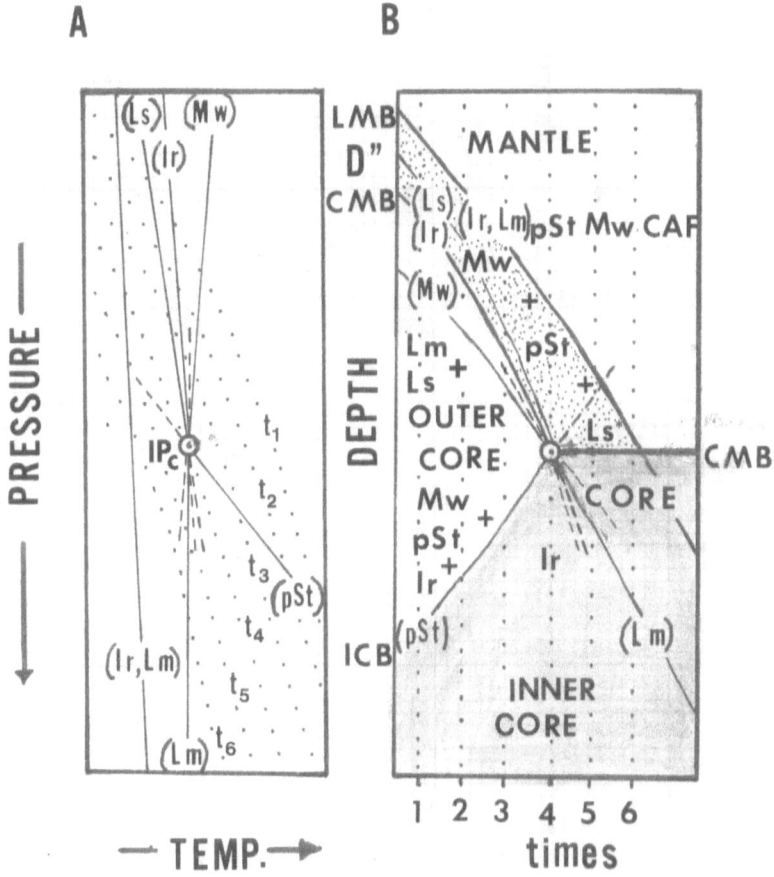

Figure 8 This illustration consists of a pressure-temperature diagram of an invariant point and univariant lines for the hypothetical ternary system, Iron-Post Stishovite-Magnesiowustite, the same as 4B and 7A, where the 'depth' for 8B corresponds ideally to pressure in 8A. This means that the intersections of six schematic cooling gradients of 8A with reactions (Ir,Lm), (Ir)and (pSt) can be transferred horizontally to 8B to create a 'six frame animated picture' of the crystallization of the Earth's outer core. Notice that the outer core in its tendency toward equilibrium lasts until 'time four' and beyond '~time six' the core-mantle system has attained a liquid absent state of equilibrium where the final CMB is a surface with crystals of pSt+Mw+CAF resting on crystalline Ir. However, this surface is no longer a univariant surface, rather it is a four phase relict of three

univariant reactions after several tens of Ga of the Earth's plate tectonic cycle, a material and energy transfer system cycling toward equilibrium, now attained. If the Earth's plate tectonic cycle began with the first formation of the Earth's inner core, because it is needed to eject immiscible silicate liquid as detailed in Fig.7, it will have certainly ended by time four to yield a solid metallic body studded with crystals of the lower mantle and some larger inclusions of crystalline aggregates of lower mantle minerals from 'frozen-in' Ls-mega-bodies. Such a structure might suggest the group of meteorites called pallasites, but these inclusions are of much greater pressure than those which allow for olivine and pyroxene that they typically contain.

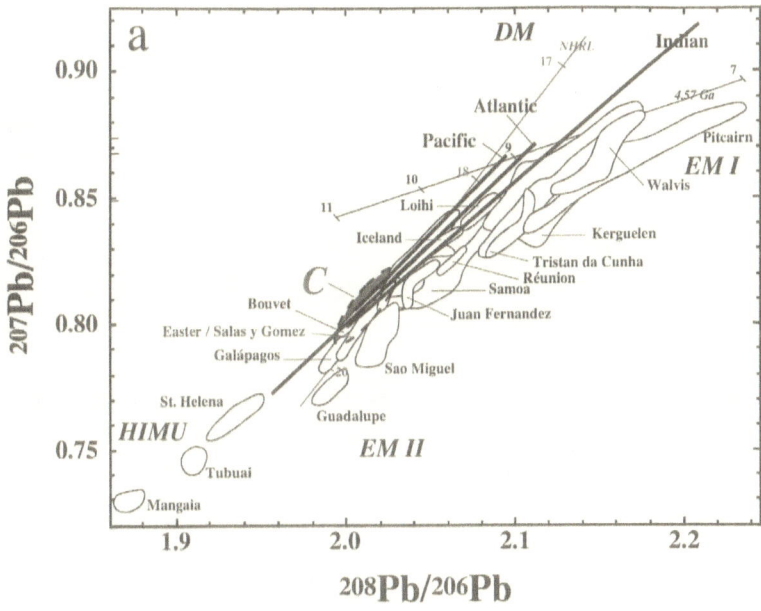

Fig. 9 is Fig 1A from Hannan and Graham SCIENCE 272:991 (17 May 1996) reprinted with permission from AAAS "^{208}Pb/^{206}Pb versus ^{207}Pb/^{206}Pb covariation diagram showing the converging regression lines for the three MORB subpopulations (Atlantic, Pacific and Indian 'ocean basins') in relation to the OIB fields and the global isotopic end-members HIMU, EM1, EM2, and DM. The end-member components encompass the entire oceanic basalt database and plot in more extreme positions than the general positions shown in the diagrams. ("We used published analyses and our own unpublished Pb and He isotope data. References for the MORB and OIB isotope database are available from Hannan and Graham on request"). One other fact that should be noted and is that the 'lines' for each of the ocean basins are defined at the 95% confidence level but they do not show the 'slight broadening' away for 'C' that is illustrated in their Fig. lB, a ternary plot of the relative abundance of ^{206}Pb, ^{207}Pb, and ^{208}PB (52).

These linear or binary arrays mentioned above, especially those of the oceanic island basalts (48), have been review by Jason P. Morgon (54) and he showed quite clearly that it is incorrect to call these linear arrays the product of the mixing of two partial melting processes, because those are demonstrably curved. Instead, he developed a model which suggests that the product of two or more sources could be expected to form a heterogeneous body characteristic for each 'hot spot'. This body is then thought to be the sequential source of basalt magmas in a 'melt extraction trajectory', so well illustrated by (48, 49,5 1, 52, 53). An alternative model suggests itself to me. If the silicate liquid ejected from the Earth's core is equivalent to C- component and FOZO, and if it is a silicate liquid of constant composition (major-, minor-, trace- and isotope-elements) in its superheated state, then when it attaches lower mantle sources it yields homogeneous products of linear variation. One could expect such products, upon delivery to our sample population, to display linear arrays with different indigenous lower mantle sources that pivot on the composition of the core's silicate liquid (= 'C-component'=FOZO) Fig. 9 in n-dimensional compositional space, which is exactly what one finds. If the 'homogeneous solid product of this superheated attach' is rich in ' ' siderophile elements, that are fractionated into the upper mantle during plume melting , inclusions of these could be a signature of this type of plume activity.

Evidence of another type of core fertilization comes from Brandon et al. (55) who showed that Os isotopes can be used to detect 0.5 to 1.0 wt. % metallic iron from the CMB or the D" layer in six recent picrite lavas from Hawaii. I contend that unknown but significant amounts of iron and metallic liquid could logically come from the Earth's core by entrainment as micro-phases of Ir or Lm(Ls) at micrometers to millimeters in size, in the ejection dynamics of mega-bodies of Ls. If openings are small enough to stop micro-phases of Ir, pSt and Mw on the core-side of the 'point of weakness' or the 'line of weakness' at the CMB due to 'filter-pressing' of Ls mega-bodies, Lm is still likely entrained because of its liquid state. The contact of pSt and Mw micro-phases at such a contact react exothermally producing Lm according to combination #79 of Table 1 to reinforce its entrainment. There are three different 'outcomes' or reaction products for the contact of pSt+Mw which includes a metallic liquid saturated with various other phases plus heat, i. e., Lm(pSt,Ls)+Ls(Lm,pSt)+pSt, or Lm(Ls)+ Ls(Lm) or Lm(Mw,Ls)+Ls(Lm,Mw)+Mw. Newly formed metallic liquid from these reactions could readily be entrained through the smallest of openings in minor but potentially significant amounts. However it hard for me to imagine egress openings small enough to 'filter-out' all micro-phases of mega-bodies of ejected silicate liquid in the first place and the Lm liquid phase would definitely pass. This model of a 'core component in Hawaiian picrites' is challenged by Scherstein et al.(56) whose tungsten isotope evidence, which used the same specimens from Hawaii as Brandon et al. (55), strongly suggests that there is no evidence that supplying plumes contain contributions from the core. They indicate radiogenic osmium in oceanic island basalts can be better explained by a source containing a component of recycled crust that included Mn-nodules! This controversy does not effect the model of the C-component equal to FOZO and their alleged origin from silicate liquid from the Earth's core, but the omission of the entrainment of Ir, and especially Lm, is not explainable by me.

According to Williams et al. (57). I suggest, herein, that ULVZs can be explained as 'wounds' on the D" layer where Ls from the outer core has been ejected, i.e., the post-ejection facies of the D" layer (Fig. 7D), rather than as the source of thermal plumes themselves. These ultra low velocity zones differ from the normal facies of the D" layer by having ~45 km of liquid rich 'zone' in their 200 km thickness, according to (58), versus only 5 km normally, according to (59). The major exception to the idea of a passive mantle is from Ritsema et al. (62). They noted that the active rifting and basaltic volcanism of the East Africa can be linked to a thermal anomaly above the CMB some 45 degrees, or 4,000 km at the surface, the so-called African supere-plume under the Southeastern Atlantic Ocean. There is, however, no explanation of why this mantle current is so large and tilted in its effect. Ni et al. (63) confirm the tilt of this large structure beneath southern Africa and they suggest that it is not stable, that is, not normally expected even on geological time scales.

Knittle and Jeanloz (64) used a direct approach in their study of the CMB by placing iron foil in contact with enstatite ($Mg_{0.88}Fe_{0.12}SiO_3$ from Bamble, Norway) and took them up to ~70 GPa at 3,500°K where enstatite converted to perovskite, a major phase of the lower mantle, and its contact with iron foil apparently resulted in spontaneous reaction in their diamond anvil device. Their products are numerous and somewhat complex: Mg Perovskite ($MgSiO_3$), Post Stishovite (SiO_2 or closely related phase [like glass? i.e., silicate liquid?]), Wustite (FeO), Iron Silicide FeSi, quenched Iron Liquid and unreached Iron. In my opinion their result can be reduced to a fundamental ternary (Fe-Si-O) reactions:

$$3Fe + SiO_2 = 2FeO + FeSi \text{ (10) plus En} = MgPv \text{ (11)}$$

If the Earth's CMB is on the Fe + SiO_2 side of reaction, which the temperature and pressure of their experiment indicates, then the relations I have presented above have a good chance of being relevant as intended. If the Earth's CMB is on the FeO + FeSi side, however, they are not and the D" layer is likely made up of Mg perovskite + FeO + FeSi + silica phase, but the silica phase must not come in contact with liquid iron of the outer core because more FeO + FeSi will result. Rather then excess silica and pure FeO though, one would expect other lower mantle phases like Mg perovskite, Ca perovskite and Magnesiowustite, (30, 32, 34) plus FeSi, to be coexisting phases in the D" and to be suspended as the four mantle-seeking phases in the outer core. They could combine to lower the density of the outer core by 10 % with an interesting combination of homogeneous and heterogeneous equilibria and they would generate energy via metallic liquid buffered convective phase contact reactions toward stable phase equilibrium, but there is no immiscible silicate liquid to 'work against the Lm body 'for convection or to be ejected from the outer core to quell its high energy state and disrupt its convection to effect a change in polarity. For these three reasons I judge this reaction as interesting but probably inapplicable to the Earth's D" layer, core and the Earth's plate tectonic cycle, and the critical reaction (10) is likely above 4,300°K at CMB pressure.

A comprehensive determination of the chemical composition of the Earth by McDonough and Sun (65) found that the Earth's mantle is depleted in Mg and Si relative to refractory lithophile elements, when compared to C1 carbonaceous chondrites. The presence of a silicate liquid component in the Earth's core proposed there at almost one and a three-fourths times the volume of the crust of the Earth (21) should bear on this problem. They indicate that there is negligible evidence for exchange between the core and the mantle for the last 3.5 Ga, and dismiss the idea that the missing Si and Mg are sequestered in the core, but that was before the unavoidable presence of silicate liquid in the Earth's core was rationalized there. Moreover, they make little of the enrichment of siderophile elements in the upper mantle by 20 to 200 times relative to that expected for equilibrium partitioning during core formation according to Urakawa (29) and Ringwood (15). Urakawa was first to discover that Ni transfers into a silicate liquid phase from an iron phase at pressures up to 170 kb pressure; a result confirmed by (27 , 28) at pressures up to 20 GPa. I suggest that this observation leads to the idea that silicate liquids ejected form the Earth's core are enriched in siderophile elements as they become hybridized in the lower mantle and ascend as basic hot plumes , through a denser and colder mantle, and that they fractionate in the upper mantle as basaltic magma of low siderophile content while enriching the upper mantle in siderophile elements. This enrichment of the upper mantle makes a case for the concept that it is not representative of the earliest mantle in its siderophile content, instead it is the sited as evidence of mantle plumes starting from the Earth's outer core that leave ponderable, secondary, siderophile element concentrations in the upper mantle.

Buffet et at. (66) propose that 'sediments' of Mg Perovskite, FeSi, 'Silica' and FeO', from the reaction of Mg,Fe Perovskite with molten Fe of the outer core reaction of (64) and reactions (10 and 11), accumulate at the top of the Earth's outer core in aniclinal irregularities to explain ultra low velocity zones, ULVZs, of the D" layer and geodetic observations, nutations. Nutations are variations in the Earth's rotations that are thought, by them, to be explained by the effects of metallic electrical conductivity in the upper 200 m of the core. There is no evidence of such mantle seeking phase (i.e., FeSi) in other relevant phase studies of this region (30, 31, 34) but the idea of anticlinal irregularities is interesting but the presence of 'Silica' there would cause additional reaction with molten iron (reaction 10), and silicate liquid of ULVZs is mainly concentrated in the bottom of the D" layer above the CMB . Since my analysis of their results (64) negates the application of their results to the Earth's plate tectonic cycle, it may apply to this hypothesis as well, if I am correct in my analysis.

Morse (67) poses an alternate view of ultra-low velocity zones (ULVZ)s. He sees them as gigantic magma chambers (GMC) of the D" layer of the Earth, created by a heat pump within the outer core driven by the crystallization of the inner core and another above the CMB that melts the mantle and continues as a heat conduit and major iron-rich volcanic expression as the heat pump takes it through the mantle to the lithosphere and crust above for each GMC, thus, explaining the connection between 'hot spots' and ULVZs (57). It is claimed that excess heat from the core occasions enough superheating to overcome the intrinsic high density of silicate liquid so that it rises to interact with mantle roof rocks to create a mantle plume that yields Fe-rich magmas and corresponding volcanos. Typically, silicate liquid at this depth is more dense than the mantle's perovskites and magnesiowustite, so superheat is necessary for turning ULVZs into thermal mantle plumes. However, there is no reason given for such a superheating event or for the restriction of silicate liquid and other lower mantle phases to ULVZs. If there are no natural restrictions for the interactions of Ls with Lm, like Ls being restricted to the mantle side of the CMB in ULVZs, and for the outer core's interactions with subducted cold lithospheric crust complexes, Fig. 6, then one arrives at the same type of phase convecting and phase interacting system as presented in Fig. 10. I have a problem with the Earth's lower core as a heat pump produced as the inner core crystallizes. If the core's crystallization is not interrupted, as proposed herein, should not the core have solidified long ago?

Rost and Revenaugh (68) discovered rigid zones 0.12 to 0.18 km thick at the top of the core, beneath the CMB and ULVZs where density considerations from seismic velocities suggest a mixture of liquid iron plus solid mantle material, density = 9.6g/cc. This finding is consistent with the idea of the minor loss of Ir and/or Lm when major amounts of 'Ls-bodies are ejected through contacts at the CMB, and is supported by Brandon et al. (55) but not by Scherstein et al. (56). If ULVZs are wounds of the D" layer as indications of Ls ejections from the Earth's core, then it is reasonable that some Ir and Lm physically entrained in Ls escapes too. So if ULVZs are liquid that is not included in the plume, as suggested for Fig. 7D, one could also have some Ir that was not included and settled back with the Ls*. Lm would not settle back because it crystallizes to Ir via the Ls absent reaction (see reaction 12 below and Fig. 4). This means there is a very dense layer where Ir is much greater than pSt+Mw at the bottom of the D" layer, equal to the high density layer found, below the liquid-rich part of the ULVZ. My interpretation of this is that, for that section of the CMB, the Ls absent reaction (12) is the CMB reaction under such ULVZs and the rigid zone produced ~79 vol. % Ir and the remainder of ~ equal volumes of pSt and Mw,

'Ls absent reaction' pSt + Mw + Ir = Lm (12)

and it should 'pinch-out' and disappear away from the ULVZ on all sides, as the CMB reaction returns to the normal Ir absent reaction' (see Fig. 4 and reaction 13)

'Ir absent reaction' pSt + Mw = Ls + Lm (13)

I believe their observation is important because it may be the 'first and only measurement for the thickness of the CMB' as 120 to 180 meters , and that could help in the mechanical analysis of the 'silicate liquid ejection problem from the CMB'. Note however, that my solution is for iron above the CMB not below it. I can think of no good explanation for a mixture of Lm plus solids below the CMB unless it could be a clot of micro-phases in Lm just below a point of weakness, but this would break-up very quickly because of the proximity of pSt and Mw together, combination 76 of Table 1, whose three spontaneous and irreversible outcomes can be summarized as Lm+Ls plus or minus pSt or Mw.

Helffrich and Kaneshima (69) analyzed the possible role of liquid immiscibility of the Fe-O-S system for the Earth's outer core. They found liquid immiscibility is common in many Iron-rich systems and suggest that liquid immiscibility is a possibility at conditions of 136 to 330 GPa and temperatures <5200°K if the concentrations of the light elements, O and S, are high enough, but they found no evidence of layering that they expected using travel times of P4KP seismic waves.

They decided that a single liquid composition of 10.5(3.5) wt.% S and 1.5 (1.5) wt. % O is compatible with wave speeds and densities throughout the outer core. The high Sulfur content agrees with the evidence of Li and Agee (27) who found it strongly fractionated into the metal liquid in their study of Ni and Co partitioning in silicate and metallic liquids. Sulfur then Oxygen would be my choices of elemental alloying elements too, after silicate liquid, as elemental density modifiers of the outer core, but these effects are only schematically referenced by my location of 'Lm' Fig 5.

CONCLUSIONS

Examination and interpretation of relevant data (Figs. 2,3)plus the use of the simple hypothetical three component system Iron (Ir)-post Stishovite (pSt)-Magnesiowustite (Mw) as a model phase diagram (Figs. 4,5,7,8), suggests a paradigm for the Earth's plate tectonic cycle (EPTC). A model that is consistent with the cycle of more energetic matter ascending from the Earth's core and less energetic material returning to it to be converted in form and energy(Figs, 1,10) for ascension. It is also a model that slows the crystallization of the Earth's inner core by alternating crystallization and melting of the Earth's inner core; in this paradigm of the EPTC the cycle starts and ends with the Earth's core, only to begin again. This means that the Earth's core is the convertor and third end member of the EPTC, a pivotal end member to the first-, the MORB end member; and a pivotal end member to the second-, the subduction end member of the EPTC(figs, 1,10). The connection between the MORB end member and the subduction end member of the plate tectonic theory by a sea-floor spreading stage (figs. 1,10), is well known, but it is incomplete. Its cyclic nature, inferred early on due to the youth of the ocean floor, must be included to make it a complete dynamic system for the whole Earth, and there must be a convertor end member to complete it!

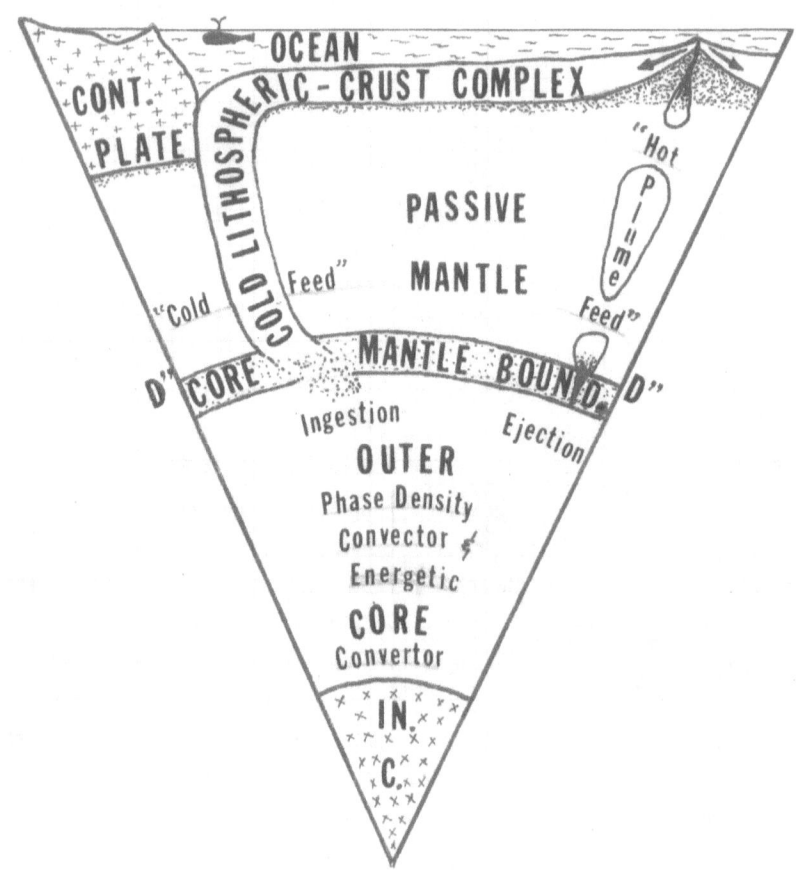

Figure 10 summarizes the Earth's Plate Tectonic Cycle in a schematic 'pie-section' of the Earth with structural and geologic features not shown in Fig. 1. It emphasizes that the Earth's core convects because of light and dense 'bodies' of the outer core that are supplied in part by the ingestion of cold lithospheric-crust complexes (CLCC)s at the D" layer, as per Fig. 6. It is also an energy convertor due to radioactivity silicate liquid and due to Ls and/or Lm buffered phase contact reactions and liquid-liquid phase contact reactions, and due to friction between liquid buffered stable contacts plus, 'gatherings and splittings' within and along shear surfaces of these light and dense immiscible mega-bodies. The amount of these so-called 'convectables' can vary independently, depending on the amounts supplied by ingestion of CLCCs. The supply of CLCCs is an independent variable because it obviously a 'top down tectonic', based on the nature of

the entire cycle and especially on the amount of interaction at shallower depths of subduction. The role of the mantle in descent of CLCCs toward the core and for the ascent of thermal plumes destine for the MORB end member of the EPTC or toward oceanic island basalts at global hot spots (not shown) is relatively passive. The best evidence for the passive role of the mantle in general is the near fixity of global hot spots and their statistically close vertical association with ultra low velocity zones

Of the crucial discoveries, which make this paradigm possible, is the discovery of Rigden et al. in 1984 (9, 10,) and verified by 11,12, 13, 14, that silicate liquid of the Earth's mantle becomes 'densified' at about 10 GPa, or about 290 km, as it 'crosses-over in density' from the least dense- to the most dense of the mantles nonmetallic phases. This cross-over means that silicate liquid could become a phase of the Earth's core. For example, 17vol.% of such a densified silicate liquid (Ls), as an immiscible liquid in a convecting iron liquid of the Earth's outer core, could lower its density by 10% at CMB conditions . If one considers the primary differentiation of a carbonaceous chondritic Earth (65), a sulfur bearing metallic melt is first to form (17) but its surface tension is too high for it to percolate through the upper mantle (18, 19), so it has been suggested that silicate melting might be necessary for the development of planetary interiors (18) like our Earth's. If so, it seems logical that this silicate liquid could release the surface tension above 290 km and relax it and follow sulfur-bearing metallic liquid below 290 km to the Earth's core. If perfect, it makes the early core at 35 vol. % of the Earth, of which 61 vol. % is probably a well stratified immiscible silicate liquid. Amazingly, this liquid likely contains significant ferric iron, from the $Fe-SiO_2-O$ system at atmospheric pressure (Fig. 2) and data on MORB-(31) and pyrolite (34) compositions at near core-mantle boundary conditions, due to reaction 7 above. Continuation of mantle convection after primary differentiation displaces this liquid upward for solidification into plumes (22) that could contribute to basic magmas in the form of huge floods of basalt and/or 'mafic intrusions'. There is little evidence remaining of these magma oceans but there are more than a few colossal mafic intrusions that are well known for their platinum group metals, nickel, chromium and other metals (25). The next stage in the depletion of this early core's silicate liquid is the asteroid bombardment of the Earth-Moon system from 3.9 to

2.8Ga. Canada's Sudbury Impact Complex (25) is a likely example but evidence for 'Earth's probable maria' are only preserved on the Moon.

When one considers the nature of the Earth's plate tectonic cycle before the last several hundreds million years, one can ask whether the rate of the cycle was more rapid in this period because of higher silicate liquid in the core? If so, does this mean that the rate of crystallization of the inner core was slower? I think the answer to both questions is yes, but the detailed answers to this 2.5 Ga span of time need much more study and are peripheral to our main concerns here that deal with the suggested current functions of the convertor end member of the Earth's plate tectonic cycle.

The first function of the convertor end member of the EPTC is to ingest cold lithospheric-crust complexes (CLCCs) from the subduction end member of the EPTC by using its heat on contact to disaggregate them by partially melting (Fig. 6). This endothermal process yields immiscible silicate liquid(Ls) and iron liquid(Lm) as it frees hypothetical lower mantle phases pSt, Mw, and core phase Ir, into the Earth's outer core (Figs. 6, 10). This process favors core cooling and crystallization of the inner core, and produces large quanities of silicate liquid and crystals because of the silicate nature of the ingesting complex, but this new product strongly interacts with passing bodies of Lm. The net result is that one should visualize the outer core as having two types of immiscible mega-bodies in convection, Ls at <17 vol. % and Lm at < 83 vol.%, both ranging in size from centimeters to kilometers. Both bodies are thought to contain significant to copious, quanities of micro-phases ranging from micrometers to millimeters in size, pSt, Mw, Ir and Ls in Lm-bodies, and pSt, Mw, Ir and Lm in Ls-bodies. It is this combination of bodies and phases in convection that reduces the bulk density of the outer core by 10 % relative to pure liquid Iron. Lm-mega-bodies would seem to traverse the full depth of the outer core with Ls-mega-bodies interacting with them so as to be mantle-seekers with a density contrast of less than two that is best expressed in their larger sizes.

Generation of heat is the second important function of the convertor end member of the EPTC and it does this by four processes. First and foremost is the heat produced by unstable contacts between and among micro-phases. These contacts react spontaneously and irreversibly to form phases of lower Gibbs free energy by exothermal processes. The second ranked source is from radioactivity of silicate liquid and perhaps other phases of the natural system (20). The third source of heat is produced by friction at stable contacts. Finally, these contact actions and reactions are important within convulsing mega-bodies and are strongly expresses along their shearing contacts and during 'gathering' processes for like bodies and during 'splitting' processes between dissimilar bodies.

The third important function of the convector end member of the EPTC is to generate the Earth's magnetic field with all of its variations of strength, focus and polarity. I assume that this can be done with the mega-bodies I have described and the heat sources mentioned, even though they are not yet quantified.

The forth important function of the Earth's core is to quell its accumulated heat which must build so as to reverse bounding univariant reactions and expand the outer core at the expense of the mantle and inner core. Reversal of the inner core boundary yields +5.0 cm^3g^{-1} for iron melting alone (76) so, it results in a overall expansion for the Earth's outer core against the passively resisting strength of the mantle. When the CMB fails , as it must, it helps to remember the shape of two flat and thin surfaces that ideally enclose an expanding sphere, i. e., a baseball for the CMB, because when it fails, it fails at points or along these 'ideal lines' of expansion, and Ls mega-bodies are explosively ejected from the outer core(Figs. 7,10). This superheated fluid under decreasing pressure is 'blown' into the lower mantle to attack its phases with great zeal at three phase boundaries and at phase surfaces until it exhausts itself as a hybrid, hot and solid domain of basic chemistry, rich in siderophile elements(27,28, 29). Its high temperature(22) and basic phase composition have sufficient density contrast (34), to ascend through a denser and colder ultrabasic mantle toward the Earth's surface. In the upper most upper mantle where plumes melt on decompression (23) they fractionate their siderophile elements into the upper mantle as molten basaltic liquids contribute new oceanic crust at the MORB end member (Fig.1) of the EPTC from 'curtain-like plumes' from line sources of the CMB . Lesser plumes from point sources on the CMB , through similar processes, add new oceanic crust at global oceanic island basalts 'Hot Spots'. Most active 'Hot Spots' are shown to have 'ultra low velocity zones' (ULVZs) below them to a very high positive degree of correlation (57), and some have discernable plumes connecting to 2800 km (60, 61), but many others do not now extend to 2800 km (60, 61). Perhaps this is so, or is it a function of the system for detecting plumes while ignoring the lower 100 km of the D" layer where ULVZs are found?

The most likely signature of the Ls so violently ejected from the outer core is considered to be the ubiquitous, yet still unexplained isotope or isotope ratio C-component in MORBasalts who's different linear variations with lower mantle sources, for each of the three major expanding ocean basins, pivot about it, as do oceanic island basalt suites about its corresponding isotope and isotope ratio FOZO component (Fig. 9). This mysterious C-component has evidence of recycling oceanic crust of 0.3 to 2.0 Ga(52) and it has the chemical complexity from continental leads as evidence of its recycled nature. I propose, as others have that, the C-component and the FOZO component are the same (52, 53) and, I contend, moreover, that they represent Ls from the Earth's outer core and , they are logically a grand average, over time, of different crustal sources that are continually added to the outer core via subduction and ingestion into the Earth's outer core, as described above. Moreover, they must also have components from the lower mantle sources which it ingests in my petrologic model of 'core to lower mantle action' for MORBasalts from the Atlantic-, Indian- and Pacific expanding ocean basins(52) and, for oceanic island basalts (48, 49, 53). And, as it may have done for a distant past, since the start of the Earth's inner core at 2.8 to 4.3 Ga ago (43, 44, 45, 46, 47). I equate that start of the inner core with the start of the EPTC because the core is essential to my mechanism for ejection of silicate liquid from the outer core, and it is a reversible action in a specific series of irreversible actions that are the Earth's Plate Tectonic Cycle (Figs. 1,10), including its ending endothermal ingestion of subducted cold lithospheric-crust complexes (Figs. 1,6, 10) that yields the 'where-with-all' for ensuing cycles.

If the Earth's plate tectonic cycle started with the beginning of the inner core at 2.8 to 4.3 Ga , its ending will be at 50 to 78 Ga according my simple extrapolations to their endings. This goal is attainable (Fig. 8) but it could be sooner if the amount of Ls in the outer core diminishes as it probably has in the past. This implies that the Earth's plate tectonic cycle is the amazing cause that 'keeps the inner core from crystallizing'. The earliest of these estimates is in agreement with the primary differentiation of the mantle (26) and Re-Os isotope evidence that suggests that the Earth's core formed very early, only several million years after accretion, and to its presence size(47)!

Many questions remain, but one more conclusion seems obvious following the work of Kesson et al. (34) on pyrolite composition that Equation 8 at 4,300°K could very well be the current CMB reaction. This is also consistent with Equations 3,5,6 form Fig. 2 and summary Equation 7 (31 , 34). The Silicate Liquid formed here contains ferric iron and would be a peritectic liquid relative to its solidus phases. As a Iron Alloy absent reaction it is one of six reactions associated with a probable invariant point in the Earth's outer core, assuming that another reaction for the inner-core boundary can be determined from these six phases, plus one, like (pSt) of the model system (Fig.4). Convection of two immiscible bodies at < 17 vol. % Silicate Liquid and < 83 vol % Iron Alloy Liquid with a density contrast of just under two, just like the model system, would seem to be the driving source for the Earth's magnetic field with its variations of strength, focus and polarity. Each body may be assumed to containing its own suits of four crystalline micro-phases plus micro-phases of Silicate Liquid in Iron Alloy Liquid and micro-phases plus Iron Alloy Liquid in Silicate Liquid. Exothermal heat comes from 'phase contact reactions', and 'radioactively' plus the actions of 'rubbings', 'splittings' and 'gathering' of mega-bodies. Binary contacts between these phases are over one hundred in three or four divisions of the outer core that results in over six hundred equilibrium outcomes like Table 1 throughout the outer core for their phase contact reactions. Their exact nature cannot be known until the relations at the invariant point are known, so much needs to be determined.

I conclude that the Earth's core is the third pivotal end member of the Earth's plate tectonic cycle--the place where the cycle starts and the place where it ends. And it ends only to become renewed again in the Earth's energy generating and convecting convertor end member, the Earth's outer core (Figs. 1,10). Further, it is unavoidable that I should suggest that certain curved features of the CMB can be thought of as a template for curved features of the MORB end member of the EPTC above, through actions of first cause as bottom-up tectonics. The subduction end member is, of course, an obvious exception because of its top-down dynamic. Furthermore, at 17 vol.% the volume of silicate liquid of the outer core is quite sizable, amounting to almost twice the volume of the entire crust of the Earth. Finally, I believe the missing source of heat may have been found to account for the heat flux of the Earth, over and above radioactive mantle sources (25)! And, I believe it has been found in this new paradigm that drives the Earth's Plate Tectonic Cycle.

REFERENCES AND NOTES

1 Hess, H. H., History of Ocean Basins, Petrologic Studies: A Volume in Honor of A. F. Buddington, Engle, H. L. James and B. F. Leonard *Editors,* The Geol. Soc. of Amer., p 599-620, 1962

2 Wegener, A., The Origin of Continents and Oceans (1929) 1966 New York: Dover Press

3 Gibbs, J. Willard, The Scientific Papers of J. Willard Gibbs **Vol. One** Thermodynamics, Longaman's Green and Co., 434p, 1906

4 Birch, F., Elasticity and Constitution of the Earth's Interior, J. Geophys. Res., **57**: 227-286, 1952

5 Birch, F., Density and Composition of Mantle and Core J. Geophys. Res., **69**: 4377-4388, 1964

6 Poirier, J. P., Light Elements in the Earth's Outer Core: A Critical Review, Earth and Planet. Intr. **85**:319-337, 1994

7 Mao, H. K., Y. Wu, L. C. Chen, J. F. Shu and J. P. Jephcoat, Static Compression of Iron to 300GPa and $Fe_{0.8}Ni_{0.2}$ Alloy to 260 GPa: Implications for Composition of the Core. J. Geophys. Res., **95**: 21737-21742, 1990

8 Dziewonski, A. M. and D. L. Anderson, Preliminary reference Earth model: Physics of the Earth and Planetary Interiors, Phys. Earth Planet. Intr. **25**: 297-356, 1981

9 Rigden, S. M., T. L. Arhens and E. M. Stolper, Densities of Liquid Silicates at High Pressures, Science **226**: 1071-1074, 1984

10 Rigden, S. M., T. L. Arhens and E. M. Stolper, J. Geophys. Res. **93**: 367-382, 1988

11 Agee, C. B. and D. Walker, Static compression and olivine flotation in ultrabasic silicate liquid, J. Geophys Res. **93**: 3437-3449, 1988

12 Agee, C. B. and D. Walker, Olivine flotation in mantle melt, Earth Planet. Sci. Lett. **114**: 315-324, 1993

13 Miller, G. M., E. M. Stolper and T. L. Arhens, The equation of state of a molten komatiite I Shock wave compression to 36 GPa, J. Geophys. Res. **96**: 11831-11848, 1991

14 Ohtani, E., Y. Nagata, A. Suzuki and T. Kato, Melting relations of peridotite and the density crossover in planetary mantles, Chem. Geol. **120**: 207-221, 1995

15 Ringwood, A. E., Chemical Evolution of the Terrestrial Planets, Geochim. Cosmochim. Acta, **30**: 41-104, 1966

16 Bowen, N. L. and F. Schairer, The System, $FeO-SiO_2$, Am. Jour. Sc., Fifth Series, **24**: 177-213, 1932

17 Agee, C. B., J. Li, M. C. Shannon and S. Cirone, Pressure-temperature phase diagram for the Allende meteorite, J. Geophys. Res. **100**: 17724-17740, 1996

18 Shannon, M. C. and C. B. Agee, High pressure constraints on percolation core formation, Geophys, Res Lett. **23**: no. 20: 2717-2720, 1996

19 Minark, W. G., F. J. Ryerson and E. B. Watson, Textural Entrapment of Core-Forming Melts, Science **272**: 530-533, 1996

20 Murthy, V. Rama, Wlm von Westrenen and Yingwei Fei, Experimental evidence that potassium is a substantial radioactive heat source in planetary cores, nature **423**: 163-165, 2003

21 Considine, D. M. and G. D. Considine Eds. Van Hostrand's Scientific Encyclopedia **Seventh Edition**, Van Hostrand Rienhold, 1990

22 Morgan, W. G. Convection Plumes in the Lower Mantle, nature **230**: 42-43, 1971

23 McKenzie, D. J., The generation and compaction of partial melts, J. Petrol. **25**: 713-765, 1984

24 Cohen, B. A., T. D. Swindle and D. A. King,, Support for the Lunar Cataclysm Hypothesis from Lumar Meteorite Impact Melt Ages, Science **290**: 1754-1756, 2000

25 Allegre, Claude, From Stone to Star: A View of Modern Geology Translated by Deborah Kurmes Van Dam, Harvard University Press, Cambridge, Mass. USA 287p, 1992

26 Bizzarro, Martin, Joel A. Baker, Henning Haack, D. Ulfbeck and Mink Rosing, Early history of Earth's crust-mantle system inferred from hafnium isotopes in chondrites, nature **421**: 931-933, 2003

27 Li, J. and C. B. Agee, Geochemistry of mantle-core differentiation at high pressure, nature **381**: 686-689, 1996

28 Thibault, Y. and M. J. Walker, Geochim. Cosmochim. Acta. **59**: 991-1002, 1994

29 Urakawa, S. Partitioning of Nickel between Magnesiowustite and Metal at High Pressure: Implications for Core Mantle Equilibrium, Earth Planet, Sci. Lett., **105**: 293-313, 1991

30 Kesson, S. E., J. D. FitzGerald and J. M. G. Shelly, Mineral chemistry and density of subducted basaltic crust at lower-mantle pressures, nature **372**: 767-769, 1994

31 ONeill, B. and Jeanloz, Abstr. EOS **74**: 584, 1993

32 Zeer, A., A. Diegeler and R. Boehler, Solidus of Earth's Deep Mantle, Science **281**: 243-246, 1998

33 Holland, K. G. and T. J. Arhens, Melting of $(Mg,Fe)_2SiO_4$ at the Core-Mantle Boundary of the Earth, Science **275**: 243-246, 1998

34 Kesson, S. E., J. D. FitzGerald and J. M. G. Shelly, Minerlogy and dynamics of a pyrolite lower mantle, nature **393**: 252-255, 1998

35 Saxena, S. K., L. S. Dubovinsky, P. Lazor, Y. Cerenius, P. Haggkvisk, M. Hanfland and J. Hu, Stability of Perovskite $(MgSiO_3)$ in the Earth's Mantle, Science **274**: 1357-1359, 1996

36 Serghiou, G., A. Zerr and R. Boehler, $(Mg,Fe)SiO_3$-Perovskite Stability Under Lower Mantle Conditions, Science **280**: 2093-2095, 1998

37 To plot wt. % analyses into the $M-MO_2-O-M_2O^*$ system, convert the analysis to mole or atom % by dividing the wt. % present by appropriate molecular- or atomic weights and calculate these fractions to 100 %. Split Al_2O_3, Fe_2O_3, etc. into 2/5 M and 3/5 O. Sum all SiO_2, TiO_2 etc. as MO_2. Sum all MgO, FeO, CaO, etc. then give half

to M and half to O. Sum Na_2O, K_2O, etc. as half MO_2 and half O. Sum all O s above plus H_2O, H_2S, CO_2, S etc. as 'O' component, gather M s as 'M' component and gather MO_2 s as 'MO_2' component. Check sum to 100 % and plot.

38 Tuttle, O. F. and N. L. Bowen, The Origin of Granite in Light of Experimental Studies in the System $KAlSi_3O_8$-$NaAlSi_3O_8$-SiO_2-H_2O. Geol. Soc America, **Memoir 74**, l63p., 1958

39 Badding,, J. V., H. K. Mao, R. J. Hemeley, in High Pressure Research: Application to the Earth and Planetary Science, Y. Syona and M. H. Manghnani, Eds. (Terrapub,Tokyo/ AGU, Washington D. C. p.363, 1999

40 Clement, Bradford M., Dependence of the duration of geomagnetic polar reversals on site latitude, nature, **428**: 637-640, 2004

41 Cox, C. M. and B. F. Chao, Detection of a Large-Scale Mass Redistribution in the Terrestrial System Since 1998, Science **297**: 831-833, 2002

42 Kendall, J. V. and P. G. Silver, Constraints from seismic anisotropy on the nature of the lowermost mantle, nature **381**: 409-412, 1996

43 Buffett, B. A., H. E. Huppert, J. R. Huppert, J. R. Lister and H. K. A. W. Woods, On the thermal evolution of the Earth, J. Geophys. Res. **101 NO. B4**: 7989-8006, 1996

44 Layer, P. W., A. Kroner and M. McWilliams, An Archean Geomaghetic Reversal in the Kaap Valley Pluton, South Africa, Science **273**:943-946, 1996

45 Glatzmaer, G. A. and P. H. Roberts, A three-dimensional self-consistent computer simulation of a geomagnetic field reversal, nature **377**: 203-209, 1995

46 Bowring, S. A. and T. Housh, Science **269**: 1535-1540, 1995

47 Meibom, A. and R. Frei, Evidence for an Ancient Osmium Isotopic Reservoir in Earth, Science **296**:516-518,2002

48 Hart, S. R., B. A. Hauri, L. A. Oshmann and J. A. Whitehead, Mantle Plumes and Entrainment: Isotope Evidence, Science **256**: 517-520, 1992

49 Hauri, E. A., J. A. Whitehead and S. R. Hart, J. Geophys. Res. **99**: 24275-24282, 1994

50 White, W. M., Geology **13**: 115, 1985: Allegre, C. J., B. Hamein, A. Proost and B. Dupre, Earth Planet. Sci. Lett. **81**: 493, 1986

51 Fraley, K. A., J. H. Natland and H. Craig, Earth Planet. Sci. Lett. **111**:183, 1992

52 Hanan, B. B. and D. W. Graham, Lead and Helium Isotope Evidence from Oceanic Basalts for a Common Deep Source of Mantle Plumes, Science **272**: 991-995, 1996

53 Hanan, Barry B., Janne Blichert-Toft, Richard Kingsley and Jean-Guy Schilling, Depleted Iceland mantle plume geochemical signature: Artifact of Multicomponent mixing?, Geochem. Geophys. Geosyst., 1 Paper. #1999GC000009, 2000

54 Morgan, J. P. Isotope Topology of Individual Hotspot Basalt Arrays: Mixing Curves or Melt Extraction Trajectories?, Geochem. Geophys. Geosyst. 1 Paper 3 1999GC000004, 1999

55 Brandon, A. D. , R. J. Walker, R. J. Morgan, M. D. Norman and H. M. Prichard, Coupled ^{186}Os and ^{187}Os Evidence for Core-Mantle Interaction, Science **280**: 1570-1573, 1998

56 Scherstein, Anders, Tim Elloitt, Chris Hawkesworth, and Marc Norman, Tungsten isotope evidence that mantle plumes contain no contribution from the Earth's core, nature, **427**: 234-237, 2004

57 Williams, Q., J. Revenaught and E. Garnero, A Correlation Between Ultra-Low Basal Velocities in the Mantle and Hot Spots, Science **281**:546-549, 1998

58 Williams, Quentin and Edward J. Garnero, Seismic Evidence for Partial Melt at the Base of Earth's, Science **273**: 1528-1530, 1996

59 Revenaugh, J. and R. Meyer, Seismic Evidence of Partial Melt Within a Possibly Ubiquitous Low-Velocity Layer at the Base of the Mantle, Science **277**: 670-673, 1997

60 Courtillot, Vincent, C. Jaupart, I. Manighetti, P. Tapponier, J. Besse, Three distinct types of hotspots in the Earth's mantle, Earth Planet. Sci. Lett., **205**: 295-308, 2003

61 Montelli, Raffaella, Guust Nolet, F. A. Dashlen, Guy Masters, E. Robert Engdahl, Shu-Huei Hung,, Finite-Frequency Tomography Reveals a Variety of Plumes in the Mantle, Science, **303**: 338-343, 2004

62 Ritsema, J., H. J. vonHeijst and J. H. Woodhouse, Complex Shear Wave Velocity Structure Imaged Beneath Africa and Iceland, Science **286**: 1925-1931, 1999

63 Ni, S., E. Tan, M. Gurnis and D. Helmberger, Sharp Sides to the African Superplume, Science **296**:1850-1846, 2002

64 Knittle, E. and R. Jeanloz, Earth's Core-Mantle Boundary Results of Experiments at High Pressures and Temperatures, Science **251**: 1438-1443, 1991

65 McDonough, W. F. and S-s Sun, The composition of the Earth, Chem. Geol. **120**: 223-253, 1995

66 Buffett, B. A., E. J. Gamero and R. Jeanloz, Sediments at the Top of Earth's Core, Science **290**: 1338-1342, 2000

67 Morse, S. A., A double magnetic heat pump at the core-mantle boundary, Am. Mineral. **85**: 1589-1594, 2000

68 Rost, S., J. Revenaugh, Seismic Detection of Rigid Zones at the Top of the Core, Science **294**: 1911-1914, 2001

69 Helffrich, George and Satoshi Kaneshima, Seismological Constraints on Core Composition from Fe-O-S Liquid Immiscibility. Science **306**: 2239-2242, 2004

70 Maun, A., J., Phase equilibria in the system $FeO-Fe_2O_3-SiO_2$: Metals **7**: 965-976, 1955

71 Darken, L. S. and R. W. Gurry, The system iron-oxygen, I The wustite field and related equilibrium, J. Am. Chem. Soc. **67**: 1398-1412, 1945

72 Greig,, J. W., E. Jensen and H. E. Merwin, Ann. Rep. Geophys. Lab Carnegie Inst., **No. 53**: 129-134, 1955

73 Washington, H. S. Chemical Analysis of Igneous Rocks U. S. Geol. Survey, Prof. Paper **No. 99** Washington, 1917

74 Tappmire, D. and J. H. Carman, Unpublished Project at the Univ. of Iowa, 1974

75 Schreinmakers, F. A. H., In-, Mono-and Divariant Equilibria, Koninki. A.k.a. Wetenschappen te Amsterdam Pro. 1915-1925, English Edition 182p reprinted 1963 The Pennsylvania State University

76 Laio, A, S. Bernard, G. L. Chiarotti, S. Scandolo and E. Tosatti, Physics of Iron at Earth's Core Conditions, Science, **287**: 1027-1030, 2000

77 I acknowledge the inspirational support of O. F. Tuttle, Richard Jahns, Peter Wyllie and Art Montana and references and encouragement came from Carl Agee, Ann B. Carman and Glen E. Carman, Barry Hanan, Dewey Moore, S. A. Morse and Hatten Yoder, Ir. Discussions with- and patience from J. R. Weidner, Ila Pluma, Dr. Earl Taitt and my former wife Carol are greatly appreciated, as is the work on my figures by William Cassidy, R. T. Weeks, and editorial suggestions by Art Montana, Hatten S. Yoder,Ir and Richard Hoppin.

TABLE 1 POSSIBLE STABLE PHASE CONTACTS OR UNSTABLE PHASE CONTACTS YIELDING ASSEMBLAGES ACCORDING TO PHASE CONTACTS AND CHANGING BULK COMPOSITION, IN ORDER OF ABUNDANCE WHERE SIGNIFICANT IN REACTANTS AND IN PRODUCTS

A WITH B A-RICH TO B-RICH

BETWEEN REACTIONS (Mw) AND (pSt) = FIG. 5

1 pSt + Ir = pSt+Ir STABLE

2 pSt + Ls(Ir,pSt) = pSt+Ls(Ir,pSt) STABLE

3 pSt + Ls(Ir,Lm) = pSt+Ls(pSt,Ir)+Ir; Ls(Ir)+Ir; Ls(Ir,Lm)

4 pSt + Ls(Lm,Mw) = pSt+Ls(pSt,Ir)+Ir; Ls(Ir)+Ir; Ls; Ls(Lm,Mw)

5 pSt + Lm(Ir,Ls) = pSt+Ir+Ls(pSt,Ir); Ir+ Ls(Ir); Ir+Lm(Ir,Ls)+Ls(Ir,Lm); Lm(Ir,Ls)

6 pSt + Lm(Mw,Ls) = pSt+Ir+Ls(pSt,Ir); Ir+ Ls(Ir);
Ir+Lm(Ir,Ls)+Ls(Ir,Lm); Lm(Ls)+Ls(Lm); Lm(Mw,Ls)

7 pSt + Lm(Mw,Ir) = pSt+Ir+Ls(pSt,Ir); Ir+Ls(Ir);
Ir+Lm(Ir,Ls)+Ls(Ir,Lm); Lm(Ls)+Ls(Lm); Lm; Lm(Mw,Ir)

8 pSt + Mw = pSt+Ls(pSt,Ir)+Ir; Ls(Ir)+Ir; Ls(Ir,Lm)+Lm(Ir,Ls)+Ir;
Ls(Lm)+Lm(Ls); Mw+Ls(Mw,Lm)+Lm(Mw,Ls); Mw

9 Mw + Ir = Mw+Ir STABLE

10 Mw + Lm(Ir,Mw) = Mw+Lm(Ir,Mw) STABLE

11 Mw + Lm(Mw,Ls) = Mw+Lm(Mw,Ls) STABLE

12 Mw + Lm(Ir,Ls) = Mw+Lm(Mw,Ls)+Ls(Mw,Lm);
Lm(Ls)+Ls(Lm); Lm(Ir,Ls,)

13 Mw + Ls(Mw,Lm) = Mw+Ls(Mw,Lm) STABLE

14 Mw + Ls(Ir,Lm) = Mw+Ls(Mw,Lm)+Lm(Mw,Ls);
Ls(Lm)+Lm(Ls); Ls(Ir,Lm)

15 Mw + Ls(Ir,pSt) = Mw+Ls(Mw,Lm)+Lm(Mw,Ls);
Ls(Lm)+Lm(Ls); Ls; Ls(Ir.pSt)

16 Ir + Ls(Ir,pSt) = Ir+Ls(Ir,pSt) STABLE

17 Ir + Ls(Ir.Lm) = Ir+Ls(Ir,Lm) STABLE

18 Ir + Ls(Mw,Lm) = Ir+Lm(Ir,Ls)+Ls(Ir,Lm); Ls(Lm)+Lm(Ls);
Ls(Mw,Lm)

19 Ir + Lm(Ir,Ls) = Ir+Lm(Ir,Ls) STABLE

20 Ir + Lm(Mw,Ls) = Ir+Lm(Ir); Lm; Lm(Mw,Ls)

21 Ir + Lm(Ir,Mw) = Ir+Lm(Ir,Mw) STABLE

22 Lm(Ir,Ls) + Ls(Ir,pSt) = Lm(Ir,Ls)+Ir+Ls(Ir,Ls); Ls(Ir)+Ir; Ls(Ir,pSt)

23 Lm(Ir,Ls) + Ls(Ir,Lm) = Lm(Ir,Ls)+Ls(Ir,Lm) STABLE

24 Lm(Ir,Ls) + Ls(Mw,Lm) = Lm(Ls)+Ls(Lm); Ls(Mw,Lm)

25 Lm(Ir,Ls) + Lm(Mw,Ls) = Lm(Ls)+Ls(Lm); Lm(Mw,Ls)

26 Lm(Ir,Ls) + Lm(Ir,Mw) = Lm; Lm(Ir,Mw)

27 Lm(Mw,Ls) + Ls(Ir,pSt) = Lm(Ls)+Ls(Lm); Ls(Ir,Lm)+Lm(Ir,Ls)+Ir; Ls(Ir)+Ir; Ls(Ir,pSt)

28 Lm(Mw,Ls) + Ls(Ir,Lm) = Lm(Ls)+Ls(Lm); Ls(Ir,Lm)

29 Lm(Mw,Ls) + Ls(Mw,Lm) = Lm(Mw,Ls)+Ls(Mw,Lm) STABLE

30 Lm(Mw,Ls) + Lm(Ir,Mw) = Lm(Mw)+Mw; Lm(Ir,Mw)

31 Lm(Ir,Mw) + Ls(Ir,pSt) = Lm(Ls)+Ls(Lm); Ls(Ir,Lm)+Lm(Ir,Ls)+Ir; Ls(Ir)+Ir: Ls(Ir,pSt)

32 Lm(Ir,Mw) + Ls(Ir,Lm) = Lm; Lm(Ls)+Ls(Lm); Ls(Ir,Lm)

33 Lm(Ir,Mw) + Ls(Lm,Mw) = Lm(Mw)+Mw; Lm(Mw,Ls)+Ls(Mw,Lm)+Mw ; Ls(Mw,Lm)

34 Ls(Ir,pSt) + Ls(Ir,Lm) = Ls(Ir)+Ir; Ls(Ir,Lm)

35 Ls(Ir,pSt) + Ls(Mw,Lm) = Ls; Ls(Mw,Lm)

36 Ls(Ir,Lm) + Ls(Mw,Lm) = Ls; Ls(Mw,Lm)

BETWEEN REACTIONS (Ir) AND (Mw)

37 pSt + Ir = pSt+Ir STABLE

38 pSt + Ls(pSt,Lm) = pSt+Ls(pSt,Lm) STABLE

39 pSt + Ls(Mw,Lm) = pSt+Ls(pSt); Ls; Ls(Mw,Lm)

40 pSt + Mw = pSt+Lm(pSt,Ls)+Ls(pSt.Lm); Ls(Lm)+Lm(Ls); Mw+Ls(Mw,Lm) +Lm (Mw,Ls); Mw

41 pSt + Lm(Ir,Mw) = pSt+Lm(pSt)+ Lm; Lm(Ir,Mw)

42 pSt + Lm(Mw,Ls) = pSt+Lm(pSt,Ls)+Ls(pSt,Lm); Lm(Ls)+Ls(Lm); Lm(Mw,Ls)

43 pSt + Lm(pSt,Ls) = pSt+Lm(pSt,Ls) STABLE

-44 pSt + Lm(pSt,Ir) = pSt+Lm(pSt,Ir) STABLE

45 Ls(pSt,Lm) + Lm(pSt,Ir) = Lm(pSt,Ls)+Ls(pSt,Lm)+pSt; Lm(pSt)+pSt; Lm(pSt,Ir)

46 Ls(pSt,Lm) + Lm(pSt,Ls) = Ls(pSt,Lm)+Lm(pSt,Ls) STABLE

47 Ls(pSt,Lm) + Lm(Mw,Ls) = Lm(Ls)+Ls(Lm); Lm(Mw,Ls)

48 Ls(pSt,Lm) + Mw = Ls(Lm)+Lm(Ls); Ls(Mw. Lm)+Mw+Lm(Mw,Ls) ; Mw

49 Ls(pSt,Lm) + Lm(Ir,Mw) = Ls(Lm)+Lm(Ls); Lm; Lm(Ir,Mw)

50 Ls(pSt,Lm) + Ir = pSt+Ls(pSt,Lm)+Lm(pSt,Ls); Lm(pSt)+pSt; Lm(pSt,Ir)+Ir+pSt; Ir

51 Ls(pSt,Lm) + Ls(Mw,Lm) = Ls(Lm)+ Lm(Ls); Ls(Mw,Lm)

52 Ls(Mw,Lm)+ Mw = Ls(Mw,Lm)+Mw STABLE

53 Ls(Mw,Lm)+ Lm(Mw,Ls) = Lm(Mw,Ls)+Ls(Mw,Lm) STABLE

54 Ls(Mw,Lm)+ Lm(Mw,Ir) = Ls(Lm)+Lm(Ls); Lm; Lm(Mw,Ir)

55 Ls(Mw,Lm)+ Lm(pSt,Ir) = Ls(Lm)+Lm(Ls); Lm(pSt,Ir)+Ir+pSt; Lm(pSt)+pSt; Lm(pSt,Ir)

56 Ls(Mw,Lm)+ Lm(Ls,pSt) = Ls(Lm)+Lm(Ls); Lm(Ls,pSt)

57 Ls(Mw,Lm) + Ir = Ls(Lm)+Lm(Ls); Lm(pSt,Ls)+Ls(pSt,Lm)+pSt; Lm(pSt) +pSt; Lm(pSt,Ir)+Ir+pSt; Ir

58 Mw + Ir = Mw+Ir STABLE

59 Mw + Lm(Ir,Mw) = Mw+Lm(Ir,Mw) STABLE

60 Mw + Lm(Mw,Ls = Mw+Lm(Mw,Ls) STABLE

61 Mw + Lm(pSt,Ls) = Mw+Lm(Mw,Ls)+Ls(Mw,Lm); Lm(Ls)+Ls(Lm); Lm(pSt,Ls)

62 Mw + Lrn(pSt,Ir) = Mw+Lm(Mw); Lm; Lm(pSt,Ir)

63 Lm(Ir,Mw) + Lm(Mw,Ls) = Lm(Mw)+Mw; Lm(Mw,Ls)

64 Lm(Ir,Mw) + Lm(pSt,Ls) = Lm; Lm(pSt,Ls).

65 Lm(Ir,Mw) + Lm(Ir,pSt) = Lm(Ir)+Ir; Lm(Ir,pSt)

66 Lm(Ir,Mw) + Ir = Lm(Ir,Mw)+Ir STABLE

67 Lm(Ir,pSt) + Lm(pSt,Ls) = Lm(pSt)+pSt; Lm(pSt,Ls)

68 Lm(Ir,pSt) + Lm(Mw,Ls) = Lm; Lm(Mw,Ls)

69 Lm(Ir,pSt) + Ir = Lm(Ir,pSt)+Ir STABLE

70 Lm(Ls,pSt) + Lm(Mw,Ls) = Lm(Ls)+Ls(Lm); Lm(Mw,Ls)

71 Lm(Ls,pSt) + Ir = Lm; Lm(Ir)+Ir; Ir

72 Lm(Mw,Ir) + Ir = Lm(Ir)+Ir ; Ir

BUFFERED BY Ls AND/OR Lm

BETWEEN REACTIONS (Ir) AND (Mw)

73 [pSt+Ls(pSt)] + [pSt+Ls(pSt)] = pSt+Ls(pSt,Lm)+Lm(pSt,Ls(pSt);
pSt+Ls(pSt)

74 [pSt+Lm(pSt)] + [Ls(Lm)+Lm(Ls)] = pSt+Ls(pSt,Lm)+Lm(pSt,Ls);
Ls(Lm)+Lm(Ls)

75 [pSt+Lm(pSt)] + [Mw+Ls(Mw)] = pSt+Ls(pSt,Lm)+Lm(pSt,Ls;
Ls(Lm)+Lm(Ls); Ls(Mw,Lm)+Mw+Lm(Mw,Ls); Mw+Ls(Mw)

76 [pSt+Lm(pSt)] + [Mw+Lm(Mw)] = pSt+Ls(pSt,Lm)+Lm(pSt,Ls);
Ls(Lm)+Lm(Ls); Ls(Mw,Lm)+Mw+Lm(Mw,Ls); Mw+Lm(Mw)

77 [pSt+Lm(pSt)] + [Ir+Lm(Ir)] = pSt+Lm(pSt,Ir)+Ir; Ir+lm(Ir)

78 [pSt+Ls(pSt)] + [Ls(Lm)+Lm(Ls)] = Ls(pSt,Lm)+pSt+Lm(pSt,Lm); Ls(Lm)+Lm(Ls)

79 [pSt+Ls(pSt)] + [Mw+Ls(Mw)] = Ls(pSt,Lm)+pSt+Lm(pSt,Lm); Ls(Lm)+Lm(Ls); Ls(Mw,Lm)+Mw+Lm(Mw.Ls); Mw+Ls(Mw)

80 [pSt+Ls(pSt)] + [Mw+Lm(Mw)] = pSt+Ls(pSt,Lm)+ Lm(pSt,Ls); Ls(Lm)+Lm(Ls); Mw+Ls(Mw,Lm)+Lm(Mw,Lm); Mw+Lm(Mw)

81 [pSt+Ls(pSt)] + [Ir+Lm(Ir)] = pSt+Ls(pSt,Lm)+Lm(pSt,Lm); pSt+Lm(pSt); Lm(pSt,Ir)+Ir+pSt; Ir+Lm(Ir)

82 [Ls(Lm)+Lm(Ls)] + [Mw+Ls(Mw)] = Ls(Mw,Lm)+Mw+Lm(Mw,Ls); Mw+Ls(Mw)

83 [Ls(Lm)+Lm(Ls)] + [Mw+Lm(Mw)]=Ls(Mw,Lm)+Mw+Lm(Mw,Ls) ; Mw+Lm(Mw)

84 [Ls(Lm)+Lm(Ls)] + [Ir+Lm(Ir)] = Ls(pSt,Lm)+Lm(pSt,Ls)+pSt; Lm(pSt)+pSt; Lm(pSt,Ir)+Ir+pSt; Ir+Lm(Ir)

85 [Lm(Ls)+Ls(Lm)] + [Ir+Lm(Ir)] = Lm; Ir+Lm(Ir)

86 [Lm(Ls)+Ls(Lm)] + [Lm(pSt)+pSt] = Lm: Lm(pSt)+pSt

87 [Lm(Ls)+Ls(Lm)] + [Lm(Mw)+Mw] = Lm(Ls,Mw)+Ls(Lm,Mw)+Mw; Lm(Mw)+Mw

88 [Mw+Ls(Mw)] + [Mw+Lm(Mw)] = Mw+Ls(Mw,Lm)+Lm(Mw,Ls); Mw+Lm(Mw)

89 [Mw+Ls(Mw)] + [Ir+Lm(Ir)] = Mw+Lm(Mw,Ls)+Ls(Mw,Lm);
Lm(Mw)+Mw; Lm(Mw,Ir)+Ir+Mw; Ir+(Lm(Ir)

90 [Mw+Lm(Mw)] + [Ir+Lm(Ir)] = Lm(Mw,Ir)+Ir+Mw; Ir+Lm(Ir)

BETWEEN REACTIONS (Mw) AND (pSt) Reference Fig. 5

91 [pSt+Ls(pSt)] + [Ir+Ls(Ir)] = pSt+Ir+Ls(pSt,Ir); Ir+Ls(Ir)

92 [pSt+Ls(pSt)] + [Ls(Lm)+Lm(Ls)] = pSt+Ir+Ls(pSt,Ir); Ls(Ir)+Ir;
Ls; Ls(Lm)+Lm(Ls)

93 [pSt+Ls(pSt)] + [Lm(Ls)+Ls(Lm)] = pSt+Ls(pSt,Ir)+Ir; Ir+Lm(Ir);
Lm(Ir,Ls)+Ir+Ls(Ir,Lm); Lm(Ls)+Ls(Lm)

94 [pSt+Ls(pSt)] + [Mw+Ls(Mw)] = pSt+Ls(pSt,Ir)+Ir;
Ls(Ir)+Ir; Ls(Ir,Lm)+Lm(Ir,Ls)+Ir; Ls(Lm)+Lm(Ls);
Ls(Mw,Lm)+Mw+Lm(Mw,Lm); Mw+Ls(Mw)

95 [pSt+Ls(pSt) + [Mw+Lm(Mw)] = pSt+Ls(pSt,Ir)+Ir;
Ls(Ir)+Ir; Ls(Ir,Lm)+Lm(Ls,Ir)+Ir; Ls(Lm)+Lm(Ls);
Ls(Mw,Lm)+Mw+Lm(Mw,Ls); Mw+Lm(Mw)

96 [pSt+Ls(pSt) + [Ir+Lm(Ir)] = pSt+Ls(pSt,Ir)+Ir; Ir+Ls(Ir);
Lm(Ir,Ls)+Ir+Ls(Ir,Lm); Ir+Lm(Ir)

97 [Ir+Ls(Ir)] + [Ls(Lm)+Lm(Ls)] = Ir+Lm(Ir,Ls)+Ls(Ir,Lm);
Ls(Lm)+Lm(Ls)

98 [Ir+Ls(Ir)] + [Lm(Ls)+Ls(Lm)] = Ir+Lm(Ir,Ls)+Ls(Ir,Lm); Lm;
Lm(Ls)+Ls(Lm)

99 [Ir+Ls(Ir)] + [Mw+Ls(Mw)] = Ir+Lm(Ir,Ls)+Ls(Ir,Lm);
Ls(Lm)+Lm(Ls); Mw+Ls(Mw,Lm)+Lm(Mw,Ls); Mw+Ls(Mw)

100 [Ir+Ls(Ir)] + [Mw+Lm(Mw)] = Ir+Lm(Ir,Ls)+Ls(Ir,Lm);
Lm(Ir)+Ir; Ir+Lm(Ir,Mw)+Mw; Mw+Lm(Mw)

101 [Ir+Ls(Ir)] + [Ir+Lm(Ir)] = Ir+Lm(Ir,Ls)+Ls(Ir,Lm); Ir+Lm(Ir)

102[Ls(Lm)+Lm(Ls)] + [Mw+Ls(Mw)] =
Ls(Mw,Lm)+Mw+Lm(Mw,Ls); Mw+Ls(Mw)

103[Ls(Lm)+Lm(Ls)] + [Mw+Lm(Mw)] =
Ls(Mw,Lm)+Mw+Lm(Mw,Ls); Mw+Lm(Mw)

104[Ls(Lm)+Lm(Ls)] + [Ir+Lm(Ir)] = Ls(Ir,Lm)+Lm(Ir,Ls)+Ir;
Ir+Lm(Ir)

105[Lm(Ls)+Ls(Lm)] + [Ir+Lm(Ir)] = Lm; Lm(Ir)+Ir

106[Lm(Ls)+Ls(Lm)] + [Lm(Mw)+Mw] = Lm; Lm(Mw)+Mw

107[Mw+Ls(Mw)] + [Mw+Lm(Mw)] =
Ls(Mw,Lm)+Mw+Lm(Mw,Ls); Mw+Lm(Mw)

108[Mw+Ls(Mw)] + [Ir+Lm(Ir)] =Mw+Ls(Mw,Lm)+Lm(Mw,Ls);
Mw+Lm(Mw); Lm(Mw,Ir)+Ir+Mw; Ir+Lm(Ir)

109[Mw+Lm(Mw)] + [Ir+Lm(Ir)] = Lm(Mw,Ir)+Ir+Mw ; Ir+Lm(Ir)

www.ingramcontent.com/pod-product-compliance
Lightning Source LLC
Chambersburg PA
CBHW021229280526
45784CB00005B/2030